# Problems & Solutions

In

# Glass Production

## Dr Eşref AYDIN

# Problems & Solutions In Glass Production
# by Eşref Aydin

Published by the Society of Glass Technology

The objects of the Society are to encourage and advance the study of the history, art, science, design, manufacture, after treatment, distribution and use of glass of any and every kind. These aims are furthered by meetings, publications, the maintenance of a library and the promotion of association with other interested persons and organisations.

Society of Glass Technology
9, Churchill Way
Chapeltown
Sheffield S32 2PY
UK

Registered Charity No. 237438

Web site: http://www.sgt.org

ISBN 978-1-917088-00-8

# Preface

Dr Eşref Aydin has been employed by Şişecam at their research laboratories in Istanbul, Turkey for many years. This company has a variety of furnaces in many different parts of the world, producing a wide range of products from flat glass to beautiful works of art. Such an environment has given Dr Aydin the opportunity to accumulate wide experience of the extensive range of problems that can occur in glass making. He has further supplemented this knowledge during a long-standing membership of the International Commission on Glass's Technical Committee on Refractories (TC11). His widely recognised experience meant that he was appointed chair for seven years. In appreciation of his expertise he has been invited to give many conference presentations; these have confirmed his ability not only to solve problems but also to pass on his knowledge to a wider audience. Such is the background that is the starting point for the present book.

His aim is to provide 'training' in the fundamentals of glass melting and this book provides an informative background to the whole range of glass making processes particularly in the Flat Glass and Container industries to a target audience which includes a newish employee within the glass fault analysis laboratory or someone whose responsibilities are a part of the process that can generate defects. It reads very clearly at a practical level, suited well to such personnel. Its scope is the whole gamut of production processes and what can go wrong. It covers the delivery, bunkering and weighing of raw materials, through glass melting, melt-refractory interactions, coating and on to post-manufacture processing such as tempering and creation of double-glazing units.

A few books already exist in this general field but perhaps too much of what has been produced previously on these topics has been unnecessarily academic and too focussed in approach; consequently, they have not been wholly accessible to a wider audience. This present contribution is therefore a valuable addition to the literature, being rich in examples and soundly based on simple principles to aid the technologist faced with an urgent problem requiring an immediate solution.

John Parker, HonFSGT, FIM, MA, PhD, Professor Emeritus at the University of Sheffield, UK

# Author's Note

This book is a final comment on what I have to say about glass production problems. The problems are numerous with small or big impact. Those which I have chosen are everyday problems and will remain at the top of the agenda. Glass is a high technology product for many applications, but a commodity product for all walks of life. It has proven itself as the most harmless product in comparison to its so-called alternatives. Glass has a long history behind it and undoubtedly has a bright future as a material.

Some problems have a long history, but technological developments eliminate most of them to satisfy customer requirements. The behaviour of glass in the environment it is used, requires special attention and care. Although glass is reckoned to be neutral and resistant, it may deteriorate in quality if necessary measures are not taken.

Daily and regular glass furnace observation is very important for best management. Energy, raw materials and labour are the main cost items in glass production. This book does not claim to cover all the fundamentals of glass. It is intended to uncover the strong conservative look that is shed on glass production since early times. Nothing remains to be secret or confidential at this time of easy access to information. I have tried to contribute to this pool of information by merging together field experience with the traditional knowledge enhanced with a touch of R&D. Many problems have been addressed and solutions proposed.

Several problems have been summarised in *Chapter 1*, raw material problems have been dealt with in *Chapter 2*; *Chapter 3* is a follow up which deals with homogeneity, batch reactions and furnace observations. Batch related problems are not by any means of lesser importance, all are tackled one by one in *Chapter 4*. Glass defects are iterated in *Chapter 5*. There is a wealth of information about defects, related origins and proposals for solutions. Tin bath and related problems are examined in *Chapter 6,* by making extensive use of available literature. Tin bath faults are dealt with in detail. Satisfactory information has been condensed about tin bath chemistry. There are some features which are completely new. Post-production problems of glass products (float, container and tableware) are given in detail in *Chapter 7.*

The reader will have noticed by scanning through the pages of this book that there is a good deal of bias on float glass. This is because I am more specialised in this aspect.

I like to think that everybody will find some titles in the contents that are of interest to the reader. The author will be indebted to those who can give some critiques and views about the book.

I humbly declare myself as an expert in glass related problems. I am proud to say that I had the best education in the world renowned SISECAM Company, I spent 32 years in R&D starting from the scratch to an internationally known department. Put in a nutshell, I can say that I learned everything here.

I feel privileged and honoured to have a 'preface' to this book prepared by eminent academician Emeritus Prof. Dr John Parker, to whom I am especially grateful.

My gratitude and sincere thanks are to everybody especially to Ertan Tanyeli who did a lot of hard work to shape up the manuscript. Thanks are also to Hüseyin Uzun and Metin Oğuz who have insistently encouraged and motivated me to write this book, and also to those who may have had a say on me during this time. I want to thank all my colleagues whom I worked in close cooperation.

Special thanks are to my family members, who motivated me during the preparation of this book.

Dr Eşref Aydın
January 2020
dr.esrefaydin@gmail.com

# CONTENTS

# CHAPTER 1

## 1.1 Main Glass Types at a Glance

Glass is reckoned to be an 'inorganic' industrial material with many favourable properties. It is due to these properties glass finds many uses in our lives. Almost all commercial glasses are produced by using natural raw materials. The chemical compositions of all glasses are designed to give specific properties, for which some specific ingredients are added or special production and forming methods are applied.

Customers are insistently pushing for higher quality. It is this never-ending interaction between the customers and the glass producers that has culminated in products of high quality. This causes a driving motivation for glass producers to improve their technologies. The major volume of commercial glasses are made up of float and container types. A typical batch is shown in Figure 1(a).

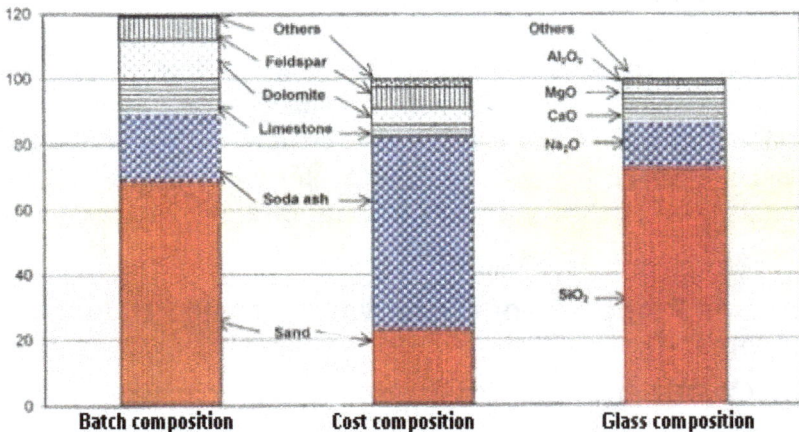

Figure 1(a): The batch, cost and oxide compositions of commercial glasses. Note that soda ash has the highest cost in glass composition.

The raw materials used in the batch recipes of commercial glasses are shown in Figure 1(a). It is possible to compare and apprehend all the glasses at a glance with respect to one another, Figure 1(b).

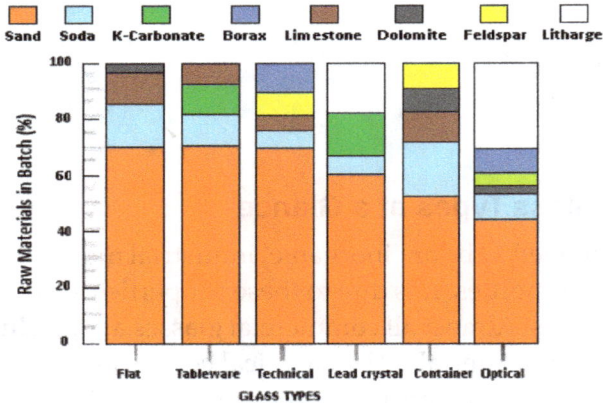

Figure 1(b): Each glass type shown in this figure should be iterated on the same basis as shown in Figure 1(a), that is each one has a batch composition as it is shown in this diagram. The cost and the oxide compositions have to be calculated separately.

Almost all oxides that make up the chemical composition of glass are obtained from rocks and minerals found in nature. All of these materials must be subjected to elaborate mineral beneficiation methods, since the minerals necessary for glass are frequently in combination with some other minerals which are detrimental and need to be separated. The minerals which can be associated with glass raw materials are given in (Appendix 1). Due to the fact that the presence of some minor ingredients affect the properties of glass, certain specifications have to be defined for raw materials. High efficiencies can be attained (e.g. 70–90%) for the production of commercial glasses, if the control of all inputs are kept live and effective.

## 1.2 Some Properties of Glass Which Create Production Problems

### 1.2.1 Colour in Glass

Colour is an important parameter which must be kept constant in all glass products. Although it is directly related to the glass composition (iron level and other colorants) colour can also be affected by the oxidation level of furnace conditions. The colour of some glasses (e.g. selenium bearing) can be strongly affected by the annealing lehr conditions. Various types of colours can be produced in all glass types, thus obtaining value added products. Some colours in some products (e.g. automotive glass) create a strategic advantage to the producer. The stability of colour strongly relies on the stability of furnace parameters. Colour is the first property which impresses the consumer.

The level of iron in glass is the major factor for colour, typical contents for $Fe_2O_3$ in the main glass types are given below. The glass user and the producer agree on a specific colour to start and maintain any consignment.

| Type of Glass | % $Fe_2O_3$ |
|---|---|
| High quality crystal glass | 0·010 |
| Handmade glass/press crystal | 0·025 |
| Clear container | 0·040 |
| Clear float glass | 0·080–0·10 |

The level of Fe in glass is the major factor for colour. The transmission value of glass at a specific wavelength is critically followed by some consumers, and yet others may be insistent on the UV transmission values of glass, like beer bottles or medical containers (amber glass). Physicochemical reactions (redox) may become important in coloured glasses. During the production of glass the redox level is maintained at a stable value, though there may be some problems (fluctuations) from time to time.

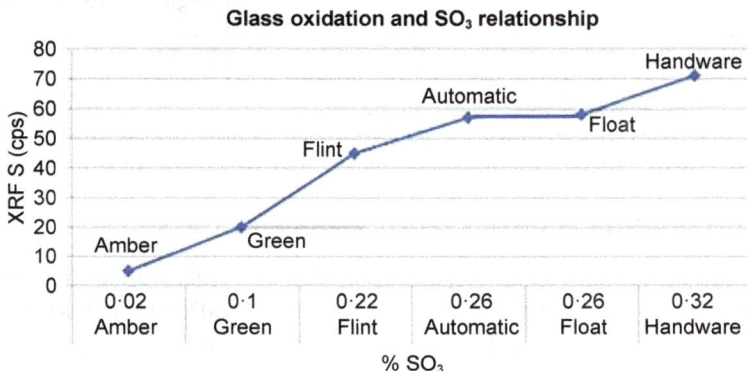

Figure 2: The $SO_3$ content (measured by XFR) of glasses is a good indicator of oxidation state. Amber glass is the most reducing, while handmade glass is the most oxidant.

The final properties of glass depend on the colour of glass, its refining state (bubble level), and on its redox state. The parameters which directly affect the redox level are $Fe^{2+}/Fe^{3+}$ ratio and the amount of $SO_3$ dissolved in glass. The $SO_3$ level dissolved in glass is a potential indicator for the redox level of that glass. The following relationship is a good guideline, Figure 2, which can be produced by XRF counts of sulphur.

The following are practical indicators for the redox level of glass:

*Very oxidising glass contains*     *> 0·30% $SO_3$*
*Very reducing glass contains*      *< 0·02% $SO_3$*

For float glass the following sequence is significant:

*Oxidising*            *Reducing*
*Grey > Bronze > Clear > Green*

$Fe^{2+}$ imparts a strong green-blue colour to the glass, while $Fe^{3+}$ gives green-yellow colour. The negative contribution of $Fe^{3+}$ can be masked or decolorized by using appropriate amounts of selenium and cobalt. The redox state of glasses can be classified into four groups, Table 1. This information is highly valuable during trouble shooting cases.

| Glass type | % $SO_3$ | %$Fe^{2+}/\sum Fe$ | Note |
|---|---|---|---|
| Oxidising clear glass (Grey, (bronze)) | 0·21–0·30 | 15–30 | Difficult to refine, $SO_3$ dissolves in glass |
| Reducing clear glass | 0·10–0·21 | 30–40 | Well refined, low seed level |
| Stable coloured glass (green, amber) | 0·04–0·06 | green: 40–55 amber: 75–82 | High $SO_3$ dissolution |
| Unstable coloured amber glass | 0·1<0·12 | 70–80 | Dark coloured glass, bubbles maybe present |

Table 1: The oxidation states of glasses are indicated taking into account their $SO_3$ and $Fe^{2+}/Fe^{3+}$ contents. Their tendency to refining is also noted.

### 1.2.2 Problems Related With Glass Composition

The similarity between float glass compositions is quite remarkable, whereas differences in container glass compositions are somehow more tolerable. On the other hand, there may be some fundamental differences in tableware and handmade glass compositions [1,2].

All glasses are made up of several oxides, which individually contribute to the characteristics of each glass. Each oxide added to the batch is designed to give some favourable properties. During the production of glass some properties have to be under continuous surveillance; these are namely viscosity, melting behaviour, crystallisation tendency and chemical resistance (durability) of glass. Some oxides are strongly effective in the positive sense, while others are equally effective in the negative sense. For understanding glass producing problems and be able to propose some solutions, it is strongly recommended that a glass technologist should have a good grasp of the information given in Table 2.

| | $SiO_2$ | $Al_2O_3$ | $Na_2O$ | $K_2O$ | $Li_2O$ | $CaO$ | $MgO$ | $BaO$ | $B_2O_3$ | $PbO$ |
|---|---|---|---|---|---|---|---|---|---|---|
| Viscosity | ++ | + | _ _ | _ _ | _ _ | O | – | – | – | – |
| Melting behaviour | _ _ | – | ++ | ++ | ++ | O | O | O | + | O |
| Crystallisation tendency | – | _ _ | ++ | ++ | ++ | ++ | + | O | – | O |
| Durability | + | ++ | – | – | – | ++ | + | O | O | + |

*Overview influences of components on some major (melt) characteristics. Legend: "++" strongly Increasing, "_ _" strongly decreasing, "O" neutral.*

Table 2: The effect of oxides on glass properties

The major oxides in float and container glass are alkali oxides and ($Na_2O$ and $K_2O$) and earth alkalies ($CaO$ and $MgO$). In float glass as $\sum R_2O$ increases there is a corresponding decrease in $Al_2O_3$ content, for example: The major oxides in float and container glass are alkali oxides

| | | | |
|---|---|---|---|
| 0·5% $Al_2O_3$ | requires | 13·698% | $\sum R_2O$ |
| 1·0% $Al_2O_3$ | requires | 13·680% | $\sum R_2O$ |
| 1·5% $Al_2O_3$ | requires | 13·674% | $\sum R_2O$ |

For a more general comparison of float and container glass compositions, use is made of the total of these oxides, by using a relationship first put forward by Dietzel.

**Dietzel Sum = ($\sum R_2O + \sum RO$)**

Some examples of float, container and tableware glass compositions are shown in the following Table 3. In the historical development of glass compositions, the main changes have been exercised on these oxides

| Sum of oxides | Clear float | Clear bottle | Green bottle | Amber | Tableware |
|---|---|---|---|---|---|
| $\sum RO$ | 12·67 | 11·36 | 11·56 | 11·30 | 10·19 |
| $\sum R_2O$ | 13·87 | 14·00 | 14·00 | 14·39 | 15·73 |
| $\sum R_2O + \sum RO$ | 26·64 | 25·36 | 25·56 | 25·69 | 25·93 |

Table 3: The alkali and earth alkali oxide contents of commercial glasses. They are formulated by Dietzel Sum.

### 1.2.3 Chemical Durability of Glass

Among several other properties, the chemical resistance (**durability**) of glass needs a special mention. It is the interaction ability of glass items with the environment in which they are used. When the composition of glass is designed initially its durability performance has to be critically iterated. It is the estimation of alkali ion migration to its environment. At high humidity the reaction rate is highly increased. There is more alkali ion migration in some glasses than others.

The durability performance of glass products has been defined as **Type I–III**, documented in **ISO-720-1985 (E)**. The higher the glass type the more alkali migration is possible. This is defined as the necessary acid (HCl) consumption for neutralisation, Table 4.

| Type of glass | Composition of glass | HCl=0·02 mol/L consumption |
|---|---|---|
| **Type I** | Neutral glass composition | <0·01 ml/g |
| **Type II** | Neutral glass surface | 0·10–0·85 ml/g |
| **Type III** | Soda–lime–silica glass composition | 0·85–1·50 ml/g |

Table 4: The chemical durability performance is classified and the limits of acid consumption are shown, according to **ISO-720-1985 (E)**.

The term 'hydrolytic resistance' is more widely used. The hydrolytic class of glass can be increased by **dealkalisation** of the surface, that is reducing the alkali level while increasing the $SiO_2$ of the surface. For instance, the surface of containers used for blood serum is upgraded from Type III to Type II by applying $SO_2$ gas into the bottles at the entrance of the annealing lehr. This creates a very important advantage by dealkalasing the surface of glass.

If there are no suitable conditions for transport, packaging, warehouse environment **glass corrosion** phenomenon can occur in all glass products. There are some precautions which can be implemented to minimize this problem.

There are some rules which are valid for all glasses. For instance, the chemical durability of any glass is closely related to the sum of alkali oxides; the higher the sum of alkalis ($Na_2O+K_2O$) the lower the chemical durability of glass. On the other hand, the increase of earth alkali oxides (CaO, MgO, and BaO) enhances the durability. Last but not least, the increase of $SiO_2$ in glass also positively affects the durability.

The presence of 1–2% $Al_2O_3$ in commercial glasses also boosts the durability. It has been firmly ascertained that 76% of float glasses produced in the world contain $Al_2O_3$ in the range 0·5–1·5% [3].

Glasses which are in the same group, e.g. float and container glass can be compared by using the criteria mentioned above. Other glasses have to be iterated case by case.

## 1.3 Melting or Dissolution Ability of Glasses

The melting or dissolution ability of glass is closely related to its composition; this fact is quantified by its viscosity behaviour. Melting and refining of glasses are much easier at higher temperatures where lower viscosity values prevail. $Log\mu=2·0$ is the viscosity value where melting is assumed to be completed, Figure 3.

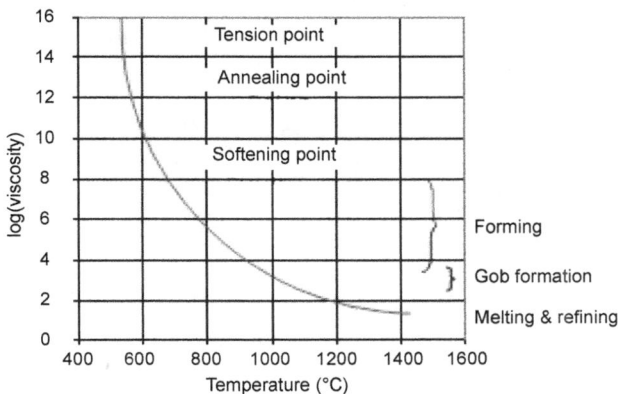

Figure 3: Viscosity of glass melt is an important parameter during glass production. Softening point of commercial glasses is between 530–700°C, while gob forming temperature is between 950–1050°C.

Changes in composition, whether intentionally or not, may give rise to some changes in the physicochemical properties of glass. The following example is intended to clarify this point, Table 5.

| Oxides (%) | Glass No: 1 | Glass No: 2 | |
|---|---|---|---|
| $SiO_2$ | 71·5 | 71·6 | |
| $Na_2O$ | 14·0 | 12·8 | |
| CaO | 10·0 | 11·2 | |
| MgO | 2·0 | 2·0 | |
| $Al_2O_3$ | 2·0 | 2·0 | |
| $SO_3$ | 0·2 | 0·2 | |
| $Fe_2O_3$ | 0·015 | 0·015 | |
| | | | Δ |
| Melting point ($log\mu=2$) | 1450°C | 1456°C | 6°C |
| Gob temp. ($log\mu=3$) | 1194°C | 1208°C | 14°C |
| Softening temp. ($log\mu=7·6$) | 731°C | 754°C | 23°C |
| Annealing point ($log\mu=13·4$) | 551°C | 575°C | 24°C |
| Working range index (WRI) | 180 | 179 | −1 |
| Relative mach. speed (RMS) | 108 | 117 | 9 |

Table 5: The effect of changing chemical composition on various parameters

If glass No: 2 has been changed for some reason, there are several and significant changes that occur. If there is a need to increase the gob temperature by 14°C, then the temperature increase in melting is about 6°C. This temperature increase in melting lowers the viscosity, which results in machine speed increase in forming [4].

### 1.3.1 Glass Defects in Glass Melting Process

Glass defects of various types and origin may occur in all stages of melting and forming. In general defects can be investigated under three groups:
- Gas containing inclusions: Bubbles, seeds and blisters
- Solid inclusions: Crystalline, cryptocrystalline or amorphous stone defects
- Vitreous inhomogeneities: Cords, cat scratches, knots, reams and striae.

The position of defect on the glass product may be an important feature. The location of the defect on the ware such as top or bottom surface, inner or outer surface, any regularity or periodicity may bear important messages for tracing the source. Whether the defects are alone or in combination with other types, e.g. stone and bubble or stone and solution zone, are important features to note. Last but not least, shape, roundness, size and crystallinity are important features to take into account.

Solid inclusions usually originate from batch related problems (incomplete transformation of quartz grains), superstructure and glass contact refractories. Segregation from batch and foam may lead to a continuous source of silica stones.

Refractory contaminants and devitrification defects may also occur. Defects may occur due to several reasons. In glass melting practice, it is a traditional approach to say that 'no changes have been made' but defects appear. To trace the reason of occurrence, one has to patiently and carefully investigate a stable period before the break out, and make necessary deductions. Basically, there is no reason why defects should occur if there are any changes in furnace parameters; this includes the raw materials as well. Stability in all aspects is an absolute must. Defects are messengers of something that is not going normal. They are like pain in the body (pain is not a disease itself, but a messenger for some abnormalities in the body).

For instance, if there are changes in temperature, cullet ratio and pull, etc., unless the new state of glass currents have reached a stable condition, there will not be a remarkable reduction in defect density. It is therefore traditional to make changes in small steps. Furnace observation at frequent intervals is necessary after any change. Furnace cameras are useful devices to follow the general behaviour of the furnace; but direct observation with glasses or even binoculars is more satisfactory whereby one can get a wealth of information. The correct management of flames with respect to glass defects originating from superstructure and glass level (metal line) is of paramount importance. It may even be necessary to apply cooling where appropriate.

If there are abnormalities in the behaviour of batch logs one must be suspicious of batch feeding problems. If the temperatures vary noticeably during the inversion the logs can lean to one side. One side of the glass surface accommodates less logs than the other side, where the logs are more crowded. The side which has less logs is subjected to a better energy transfer, resulting in a better bubble release. As the logs lean to the metal line on the other side, there is a sweeping action of logs on the refractory; which may even cause some erosion. In this way, it is highly possible that some refractory defects will be released from the metal line. These will give stones and knots.

On the other hand, if the batch homogeneity is far from ideal the transformation of sand grains (quartz) to tridymite may be interrupted, thus giving some batch defect.

## 1.3.2 Vitreous Defects: Never-ending Problem in Glass Production

Vitreous defects occur most frequently in all glass production types. Not only do they affect the quality of glass items, but they also aesthetically spoil the appearance. These defects are generally known as **knots and cords** in all production types, but in container and tableware production a generic name of '**cat scratches** 'is also equally used [5–10]. The appearance resembles 'cat claw scratches'.

The chemical composition of vitreous defects are different than the hosting glass. About 80–90% of cat scratches originate from the AZS refractories in the melter, and are characterised by the presence of $ZrO_2$ (<5%) and $Al_2O_3$ (<25%). Some cat scratches (10–20%) may originate from working end and forehearths. These cords are enriched in $Al_2O_3$ and do not contain $ZrO_2$. The source is aluminosilicate refractory types used in this part of the furnace. In general, high temperatures aggravates the problem. Under suitable conditions the influx of cords may be continuous.

Very often it is possible to find knots of very small size in a cord, where secondary crystal phases (zirconia or secondary alumina). Because of the resemblance with cords in flat glass, the term '**ream** 'is misused. Ream is two dimensional in shape while, cord is more or less one dimensional. Ream is a special and serious defect in flat glass, which is widely assumed to be a problem of chemical inhomogeneity of the glass melt with a characteristic $SiO_2$ enrichment of 0·5–1·0%. The precautions for ream are different in many ways.

The following thought or assumption has been adopted since many decades in glass production that '**vitreous *defects (cords and knots) are heavier than the base glass,* and *therefore sink to the bottom*'.** This thinking may have arisen because of the presence of $ZrO_2$ in cord composition. Based on a detailed research which we carried out over several years, it was established that it was the extent of $Al_2O_3$ content in cords which affects density. Higher $Al_2O_3$ bearing cords have smaller density than the base glass. Several melting experiments were made to verify this point. The very small magnitude of $ZrO_2$ (<5%) has minimal effect on cord density. It is due to this fact that, we put forward this interesting finding that *'cords are not of heavier density than glass; or in other words cords are lighter than glass and do not sink to the bottom of glass'*. This is absolutely the opposite of what is classically known. It is a revolutionary thought about vitreous defects in glass production. This view has been communicated across several glass platforms since 2004. Hopefully, one day it will be accepted by the glass community and the whole issue will be reinterpreted. The density of cords is determined to be 5–10% lighter than soda–lime–silica glass. *Cords do not sink to the bottom, but they travel near the top of glass melt and are carried forward by the glass current.*

Because of the fact that the vitreous defect making material does not sink to the bottom some of the classical precautionary measures such as stirrers and dip drains will have to be reiterated. In our experience it has been clearly ascertained that dip drains do not have any positive contribution; and classically stirrers pumping from bottom to top have minimal affect in eliminating cords. The stirrers should pump from top to bottom for a better result, according to the new view.

In addition to the above measures, there are more steps to be taken for eliminating or minimising cords in the source. Assuming that good quality refractories are used in the superstructure of the furnace, the quality seed and bubble level should clearly be monitored, so that low enough suitable temperatures are used to minimize refractory corrosion. Direct flame impingement on refractories should be avoided. The carryover of fine grained raw material also aggravates the refractory corrosion. Excessive insulation does not help the cord problem.

The drops and the rundowns seen at port arches and tuck stones at the glass level are potential candidates for cords and knots (one drop of AZS rundown yields approximately 400 knots). The rundowns may take the form of continuous curtain flow (due to lower viscosity) at or near the hot point, which is of higher risk. The presence of rundowns and curtain flow can be found in all glass furnaces; therefore the problem gains a never ending nature. So long as these are present in glass furnaces, the problem seems to be of a continuous character.

### 1.3.3 Gaseous Inclusions (Seeds and Bubbles)

The melting and dissolution of glass raw materials yield a consider-able amount of gases (mainly carbonates, sulphates and nitrates). Some of these gases can be dissolved in glass, while the majority are released from the glass melt and ascending to the surface according to Stoke's Law. The diameter of bubbles increases as they ascend, this is known as **fining**; on the other hand, some of the bubbles decrease in diameter and disappear as they dissolve in the glass. This phenomenon is known as **refining**. Bubbles and seeds can be present in all glass types. The number and the size of bubbles that can be tolerated by the customer can vary according to the end use, for which they are very insistent. This defect is a quality parameter for the glass item. In fact, bubble level is a pull limiting factor for the furnace. To maintain an acceptable bubble or seed level in the glass type, exhaustive discussions are made during the design stage of the furnace. In this respect, the length and width of the furnace along with the glass depth are critical parameters to be firmly decided. These studies are made with mathematical modelling programmes. All these are necessary for cost optimisation of the investment.

Each furnace has its own background characteristics for bubble potential; and which furnace operation parameters are critical for bubble are known to the furnace operators. However in real life, some minor fluctuations in some of the furnace parameters (pull, temperature, cullet ratio, batch related problems and oxidation level) may give rise to a long lasting bubble issue [11].

Specialised glass technologist most of the times can unravel the problem in cooperation with the furnace operators. After determin-ing satisfactorily the size, shape and the distribution on the ware the bubbles can be analysed to determine the gas content by using either mass spectrometer or gas chromatograph. The gas species ($N_2$, $CO_2$, CO, $O_2$, $SO_2$, Ar, $H_2$, $H_2S$, COS) can be quantified by employing any of these analytical techniques.

The analyses are not sufficient alone to solve the problem. The in-formation must be shared with the furnace operators, so that suitable action can be implemented. An internal pressure value is provided by the analysis, which is an important indicator for the bubble source. Low pressure bubbles are characteristic of the melter; while high pressure bubbles originate anywhere after the working end and the forehearth canals. In our experience six types of bubble sources have been differentiated in float, container and tableware glasses [12].

- **Melting – Refining**: If the conditions for melting and refining are not suitable, bubbles of low internal pressure can occur with the following chemical characteristics ($CO_2 > N_2$, approx. 3:1 ratio).
- **Reboil**: Bubbles which appear after a temperature rise ($SO_2 > CO_2$, approx. 15:1 ratio).
- **Devitrification**: Crystalline glass may remelt after temperature rise ($N_2 > CO_2$, approx. 2:1 ratio).
- **Contamination**: Metallic contaminants in the batch may give rise to specific type of bubbles ($CO_2 > H_2 > N_2$, approx. 10:2:1 ratio).
- **Electrolysis**: Glass gains electrolytic properties above 600°C. If under these circumstances DC voltage is applied to glass, oxygen can be released to give bubbles ($O_2 > N_2$, approx. 90:1).
- **Air entrapment**: During the movement of glass melt or by stirring action air can be entrapped into the glass to give bubbles ($N_2 > CO_2$, approx. 15:1, and up to about 1% argon is characteristic).

The chemistry of gas content of bubble changes progressively until the glass is cold enough. Due to this fact the interpretation of gas analyses must be made carefully.

The terminology used, by ICG, for gaseous inclusions is determined by the size (seed<0·1 mm, bubble<0·5 mm, blister>1 mm). These values are critical for glass for everyday evaluations and also for customer as quality criteria.


## 1.4 Problems Related with Cullet

Cullet is an important raw material in glass production, its use may reach 60–80% in Europe. It has positive contributions for melting; however if it is above 40% the time of refining of the glass may increase.

Cullet is used instead of raw materials on equivalent bases, therefore raw materials from nature are saved. On the other hand, the use of cullet saves energy, and as a consequence $CO_2$ emissions are reduced. Soda ash reacts with cullet in the early stage of batch reactions, thus consuming some of the soda, because of this fact there may be some delay in soda–sand reactions. Unfortunately, cullet is assumed to be rather pure in the batch calculations. The reality is that this is not the case; cullet is always contaminated to some extent (foreign cullet, organics, glass ceramics, metals, etc). Usually cullet is used between 25–40% in the batch; it is unbelievable to see that cullet does not receive the necessary attention analytically. This is probably due to the fact that there is no standardised analytical method, and because it is a tedious job. Frequently cullet may be the source of several problems because of contaminants and varying chemical composition. Some problems related to cullet are listed below:

- In container production the cullet used is susceptible to various contaminants, the presence of glass-ceramics are of special interest because they are transparent and cannot be noticed before use. In the furnace they do not dissolve and give stones, the smaller ones may give knots with some minute crystals. They can be investigated by using a polarising microscope or electron microprobe. If the use of contaminated cullet is discontinued it may take 3–4 days for the defects to vanish.
- Metallic aluminium in cullet gives rise to metallic silicon and many bubbles.

Refractory contaminants may get into the cullet in the plant site if necessary precautions are not taken. These are a source for stone defects.

### 1.4.1 Problems Related with Batch and Raw Materials

The melting rate of batch in a glass furnace depends on several factors. The energy required for melting is an important cost parameter, but the quality of glass should top the list of priorities. In other words, first glass quality and then energy. Some points of importance are:

- Glass composition,
- Glass raw materials,
- Granulometric properties (grain size)
- Batch feeding,
- Furnace design.

The main purpose of glass melting process is to completely dissolve the most refractory component, quartz (melting point 1740°C), in a continuous manner. Some effects of grain size are itemised below:

- Coarse sand dissolves more slowly than fine grained sand. Fine sand reacts with soda ash more quickly than coarse grained sand, which produces a sodium silicate initially. This development consumes some of the soda and the dissolution of coarse grains becomes more difficult. This is one of the main reasons why raw materials are required to have very small fine grain portion. The fine portion of the batch may easily be carried over to the regenerators if the angle of flames is not suitable, this is another reason why the fine fractions are not preferred.

- Fine grained raw material has some other disadvantages. Because of the entrainment of air into fine grained raw material, the fining of glass may become highly difficult. This forms a limitation for quality.

- During the grinding stages of raw materials the higher proportion of fine fraction is where $Fe_2O_3$ is concentrated. To eliminate the fine portion is one way of reducing $Fe_2O_3$.

- Another important problem related with the raw materials is the increase of $Fe_2O_3$, very often related to geological setting, affect the glass temperature and colour of the glass. Mineralogic investigation is necessary to decide which Fe-bearing minerals are present, so that necessary precautions can be taken by the geologists or the ore processing unit. An effective method to separate iron bearing minerals is magnetic separation either dry or in wet medium. Table 6 shows some of the common Fe-bearing minerals and the intensity range (Gauss) that is required. The minerals used for colouring the glass must be very fine grained (<0·1 mm).

| $Fe_2O_3$ bearing minerals | Gauss |
|---|---|
| Magnetite | 200 |
| Ilmenite | 2500–3100 |
| Chromite | 3900–4700 |
| Biotite | 4500–5100 |
| Pyroxene | 4800–6600 |
| Tourmaline | 4700–7200 |

Table 6: Fe-bearing minerals frequently found in glass raw materials and their best recovery range.

The criteria for selection of raw materials depend on the type of glass, and the $Fe_2O_3$ content it has. In general, raw materials are preferred to be less than (<1·0 mm), but there is a range of 0·2–0·3 mm in which sand is preferred to be concentrated, with the <0·1 mm with a possible minimum (max. 2%).

Due to the low decomposition temperatures (<900°C) of carbon-ates (limestone and dolomite) slightly coarser grain size can be tolerated, if they do not contain coarse grained free quartz above 0·5 mm. Any size coarser than this size is liable to give silica stones in the production.

- Heavy soda ash is preferred in glass production with a grain size be concentrated in the range 0·3–0·4 mm. Although, the presence of minor amount of chloride in soda ash and sodium sulphate may accelerate melting, this component has negative effect on refrac-tory corrosion and causes serious emissions. The chloride limit in these two raw materials is tentatively pushed well below to <0·2%.

Frequently different materials are crushed and ground in the same system. The grinding media in such systems (metal, alumina, porce-lain or silex) may crumble into fragments which are mixed with the raw material, giving stone defects into the glass melt. These types of problems can be unravelled by mineralogic investigation.

If there is a break out of defects related to raw materials there are some points to take into account. There is about three times of the pull molten glass in a float furnace, 1·5–2 times in a container and tableware furnaces. This amount of molten glass will be contaminated with the defect in question. Depending on the amount of cullet or raw material used and the density of glass defect in production, it may take 3–4 days to clear off. The grain size of some raw materials may be critical. For instance, if calumite (furnace slag) is used instead of feldspar, there may be some alumina fragments, which are difficult to dissolve. Even though feldspar itself is a flux material, if the grain size exceeds 0·8 mm, then some aluminous knots may appear. This is because of the high viscosity imparted by alumina rich portions in the melt, which may not be homogenised well enough. The type of defect which may disturb the production, from time to time, are roughly known by the furnace operators. This is why there is an alert situation all the time; in case there may be any unusual and unexpected defect break out.

There have been various defect break outs in the authors ex-perience, therefore we had a chance to build a dynamic 'Trouble Shooting Systematics'. In the task force there are members from furnace operators and glass technologists of various disciplines. Brain storming about the problem proves to be quite useful to solve the question as soon as possible. This motivates all the people who take part in the study. In trouble shooting studies the first proposal does not have to be 100% correct; proposals above 60–70% confidence level should be put to practice in a sequence that is appropriate. After this the confidence level will increase. It is good practice to build a Pareto Diagram which contains all the possibilities.

Mixing or misweighing of raw materials is quite frequent. The analytical approach to be adopted in these type of cases are shown in table form (Appendix 2). In any glass production line if there is a break out of defects, raw materials and the batching system are the first to be blamed, therefore special expertise should be developed for these type of problems.

Batch humidity is generally accepted to be 3–4·5%. To avoid segregation of raw materials during transport, only 2% humidity is enough, the rest is used to accelerate batch reactions. Other than humidity, batch temperature is also an important factor. The temperature of the batch in the bunker should never be less than 32°C. Below this temperature soda ash ($Na_2CO_3$) reacts with water to form $Na_2CO_3.10H_2O$, in this case there will not be free water available in the batch, therefore it will be dry, which will accelerate dusting of the batch. On the other hand, excess water means more energy. Research made on this topic indicates that 1% of humidity is equivalent to an additional 0·5% energy consumption.

### 1.4.2 Contaminants of Unknown Origin

Quite often the origin of the problem may not be related with raw materials, in this case one must look for other possibilities. The investigations carried out for each raw material are different, therefore appropriate methods have to be put into practice. The mixing of contaminants into the batch of raw materials may happen in various ways. The most frequently encountered contamination results from the transport vehicles (trucks, wagons, vessels, etc).

They may have carried some critical material before. This seems normal at first, but in reality these vehicles must be cleaned extremely well, before any glass raw material is loaded. Everybody in the chain of glass production or any form of logistics must share the same understanding and concern about glass quality. The following example explains the importance of this point. Assume that there is a truck load of 20 tons of glass sand contaminated with 1 gram of 1 mm size chromite grains. That means about 420 pieces of chromite, which do not have any chance of dissolution in the glass melt. This is equivalent to 21 particles in one ton of sand. One ton of sand produces 100 m² of 4 mm thick glass. In other words there are two defects per 10 m². The clear message of this example is that this glass cannot be sold. In this scenario, it is clearly shown that the glass producer has suffered very big losses. This crazy situation must be clearly explained to the truck owner and transport company.

| Raw material | $Cr_2O_3$ contamination |
|---|---|
| Sand | 10 ppm |
| Soda | 30 ppm |
| Limestone | 50 ppm |
| Dolomite | 70 ppm |
| Albite | 60 ppm |
| Na-sulphate | 1800 ppm |

Table 7: The chromite contamination content of glass raw materials which can cause 5 ppm $Cr_2O_3$ in glass.

Fine grained chromite itself is used as a colourant in glass to impart green colour. For a contamination level of 5 ppm $Cr_2O_3$ in glass the $Cr_2O_3$ level of each raw material is shown in the following Table 7. These are calculated on the basis they are found in the batch.

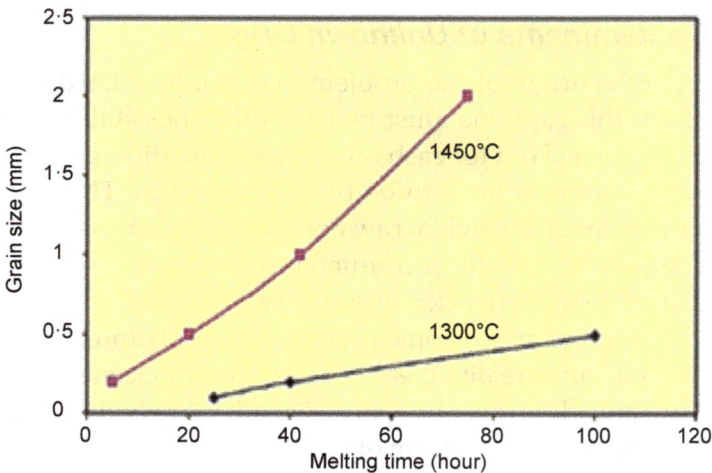

Figure 4: The relationship of grain size of chromite and temperature and dissolution time.

Very fine chromite may negatively affect the colour of glass, coarse grains give potential stone defect. In float glass the critical size for chromite is 200 μm, anything coarser than this cannot dissolve in the glass melt. Sufficient experience has been gained for the behaviour of chromite through the years.

The relationship between chromite grain size and the temperature is shown in the following Figure 4. The intermediate values can be extrapolated using the figure.

## 1.5 Systematic Analytical Approach for Glass Defects in Glass Products

It is important to have a preliminary knowledge about the defect intensity in the product under investigation (float, container and tableware, etc). At first step it is essential to collect statistically sufficient samples for investigation.

For glasses produced the number of samples to be collected and for raw materials the amounts to be investigated are indicated in Table 8.

| Type of product | Min. number of samples | Sampling procedure |
|---|---|---|
| Float glass | 25–50 defect, defect/10 m$^2$ | Samples marked by the sensors, at least 2–3 shifts |
| Container glass | 25–50 defect, % defect | Samples collected by QC personnel for 2–3 shifts |
| Tableware glass | 25–50 defect, % defect | Samples collected by QC personnel for 2–3 shifts |

| Raw material | Sample quantity | Sampling procedure |
|---|---|---|
| Sand | 10 kg | From normal sand and from problematic part |
| Dolomite Limestone | Dol. 5 kg, Lime. 1 kg | From normal product and from problematic part |
| Feldspar | 2 kg | From normal product and from problematic part |
| Soda ash, sulphate | 2 kg | From normal product and from problematic part |

Table 8: Sampling of defects in glass products (top) and glass raw materials (bottom). These are statistically sufficient to obtain reliable result.

**Special Note:** *This book does not attempt to give any details about the analytical techniques used in material investigation. Access to such information is possible in various ways. What is more important is that it will be sufficient to know where these are available, whether in house, a private laboratory or universities. For example, there is no need to go into details of mineralogic analyses, colour determination of glass and bubble analyses. It will be sufficient to know how to access these facilities.*

# CHAPTER 2

## 2.1 Problems Encountered in Glass Raw Materials
## General Information

Almost all raw materials are obtained from nature. The glass raw materials are brought together, according to a recipe, which indicates the proportion of each raw material. These are carefully mixed and homogenised in sophisticated mixers, after which this mixture is charged into the furnace. The aim is to melt this batch at temperatures of 1500–1600°C and obtain the glass itself. This melt is pulled from the furnace under decreasing temperatures, during which there is no time for crystallisation, thus forming a transparent material, glass, at the end.

Float and container glass production make up the largest volume in commercial glasses. About 98% of the constituents are made up of sand, soda ash, limestone and dolomite. The remaining 2% is Na-sulphate, a refining agent, feldspar to introduce $Al_2O_3$ into the glass, for improving chemical resistance, and any colouring agent for coloured glass production, and water content to prevent batch segregation. In addition to the above materials cullet must be added to the batch (15–25% in float glass, and 20–40% in container glass). Excluding cullet, sand forms 62–65% by weight of the batch, but the cost of sand is approximately 23–25%. On the other hand, soda ash which forms 16–20% by weight, is the most expensive (60%) component of the batch (Figure 1). The proportions of each raw material in the recipe and their cost contribution is indicated in Figure 1. In order to prepare the raw materials to satisfy the specifications by the glass producer, there are several costly steps in the mining activities, which start with the geological exploration. The cost of raw materials makes up 20% of the total cost of glass production [13]. As shown in Figure 5 there are several parameters which make up the cost, but energy stands out as number one (28%).

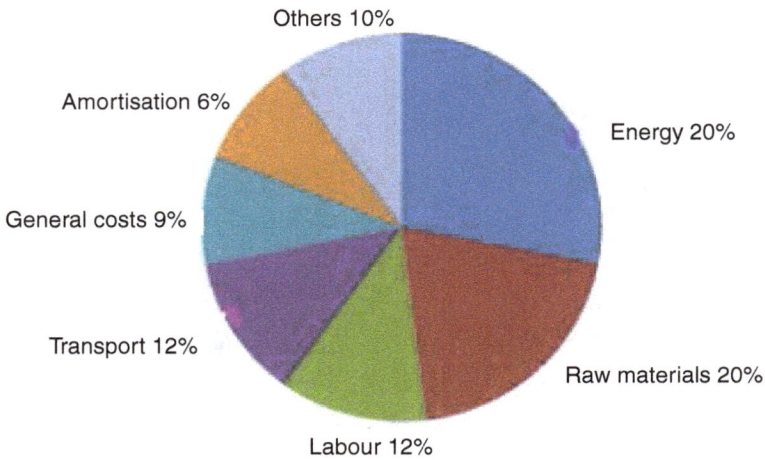

Figure 5: Glass production parameters and their cost contribution. Raw materials (20%) and energy (28%) forms the lions share.

The basic requirement for raw materials is that the specification should be adhered to all the time, and should be delivered in time to the glass production site, with all the necessary precautions taken for the logistics. Two points for the raw materials are critical; the stability of the chemical composition and the physical properties. There should be no tolerance by the glass producer. If some changes are unavoidable the supplier and the user must inform one another, well before the change to avoid any crisis. Cooperation of this nature is vital for problem free production. If changes are made at short notice, problems are likely to occur.

When a raw material is not delivered on time, or a contaminated raw material transported to the glass plant, serious problems may undoubtedly happen.

## 2.2 Problems Which May Occur in Glass Raw Materials

### 2.2.1 The Effect of Fluctuations of Physical and Chemical Properties on Melting

It may be difficult to predict any changes in raw materials. For example, fluctuations of very small amounts of colouring ingredients (Co, Mn, Cr, etc) in the quarry due to the geological character of the deposit, which may create some problems related with the colour of

glass. This may even affect furnace bottom temperatures. This type
of occurrence may take some time to solve. Sometimes colouring
agents of this type may not be related with the geology of the deposit,
but they may well be associated with the metallic components used
in the ore beneficiation system. Erosion or chemical reaction could
quite conceivably take place in the long run and causing continuing
problems. The choice of metals at the project state is critical, and
therefore developments applied to the system, on good will, should
be shared and discussed with the glass producer. If there is a noticeable progress both the supplier and the user can be motivated. If
one side is plunged into big losses, the company will face a big risk,
which will directly affect the other side. Therefore, there should be
a mutual win–win understanding.

If there is no plausible reason the chemical composition of glass
and therefore raw materials should not be changed. Should there be
any change the colour of glass and the chemical resistance and some
other physical properties may dually be changed. If so, the glass will
be a different product with new properties. This may cause some
losses in the efficiency of production, changes in the furnace regime
and fluctuations in emission levels.

The basic requirement in glass production is the necessity of stability all the time, with no tolerance. The only way to maintain this
is to stringently control all the raw materials. If the input (batch)
is stable it is quite likely that the output (glass) will also be stable.
As glass of high quality is pulled for forming operations from the
furnace, an equivalent amount of batch is charged into furnace for
melting. The glass which is pulled from the furnace at any time has
the imprints of many batch charges, all homogenised in the furnace
to give the required glass.

There is about 1500–1700 tons of molten glass on average in a float
furnace (about three times the pull), the corresponding figure for
container (200–400 tons and for tableware 100–200 tons of molten
glass, respectively, about 1·5 times of the pull). This means that this
amount of glass has to be pulled from the furnace while an equivalent
amount of batch is charged.

**Residence time** is the time the glass remains in the furnace until
it attains the quality required. It varies from furnace to furnace and
also with pull. Typical values for residence time in float furnaces

are 70–80 h, while container and tableware furnaces have values between 25–30 h. Residence time can be calculated with a high level of confidence, with the following relationship:

*(Fusion of glass in furnace/pull of furnace)×0·9×24*

Once the batch is charged into the furnace the glass formed is of very low quality, on the other hand the glass pulled from the furnace at the same time, having completed the residence time is of high quality.

It is highly important to understand the implications of the sequence of events, especially if there is an upset in the furnace. The first signs of reaction of the furnace due to some changes show themselves at or near the residence time. The furnace operators are well aware of this fact, and must show the patience necessary. One must be careful in float furnaces; there may be a straightforward flow sometime if there is some disturbance in glass currents, in this case an 8–10 h residence time is quite possible but this is not a normal situation. It is obvious that any changes made in the batch have to be given enough time. Any sporadic changes should not be allowed. This is why glass production is widely reckoned to be rather conservative.

If the chemical composition of two raw materials are very similar, replacement of one with the other may be possible, but some of the physical properties of raw materials are likely to differ. For example if there are some differences in grain size distribution, even though the chemical composition is similar, some noticeable differences, like refining ability, may occur in the melting behaviour. Under these circumstances, it is ideal for the glass producer to increase the analytical frequencies for better control. Both the glass producer and the raw material supplier must have adequate analytical facilities. To solve the problem at the source, it is preferable that the supplier has more control in the sense of analyses.

### 2.2.2 Variations in Natural Raw Materials

The geologic properties of sand may differ, both laterally and vertically, to varying scales, changes in chemical composition and granulometric aspect. The sand deposits cannot be used as they are, they must be mixed and homogenised to meet the requirements of the glass producer.

Some fossil bearing layers may frequently be present, if they are not eliminated the CaO and MgO contents may vary. Similarly, clay bearing layers may also be present. Clay, which may cause variations in $Al_2O_3$, can be eliminated at the stage of scrubbing and classification. Coal seams, manganese and phosphate nodules frequently occur in sand deposits. Of course, the main component which is prone to variation is the iron content of the deposit.

It is the primary duty of the mining section to take care of this, by applying all means available. In all these stages detailed mineralogic investigation must be applied.

The iron bearing minerals must be identified, it is also important to know whether limonite is a regular component of quartz grains, Figure 6. The increase of fine fraction in sand is an obvious reason for iron to increase, therefore this must be minimised.

This is a good reason for insisting on raw materials with the minimum amount of fine fraction possible. The solutions for the above mentioned differences are all within the professional flexibility of the mineral processing engineer.

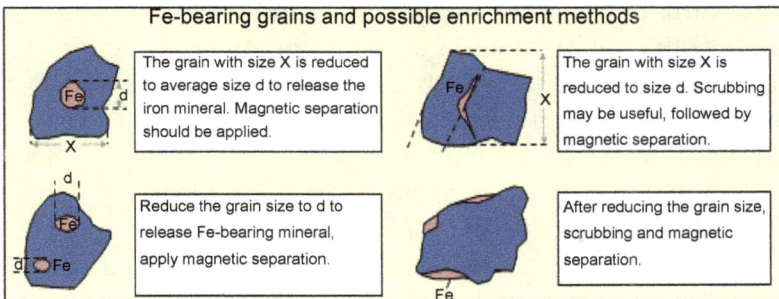

Figure 6: Quartz grains may contain limonite which may create some difficulties in the enrichment process.

The sand delivered by the supplier is discharged into silos or covered stock areas in the glass plant. Some physical or even chemical changes may occur at this stage. Segregation of fine grains from coarse grains during pneumatic charging into silos is quite frequent [14–17], which results in some type of layering, Figure 7. The fine grained portion is accumulated at the top of the silo.

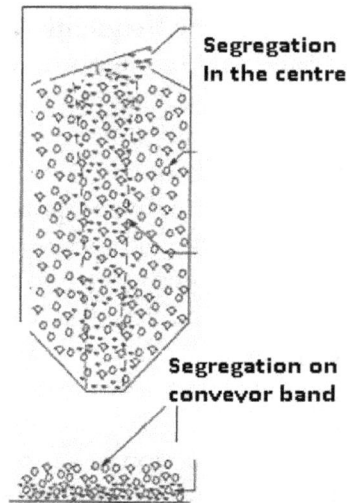

Figure 7: The grain size distribution may create some problems, the fines show an enrichment on the top of silos. This is why the input and output grain sizes may differ.

This variation may lead to physical and chemical changes. As a result of this differentiation in the granulometric properties, the newly charged material and that of discharged from the silo may show considerable differences.

The granulometric distribution of the raw material may affect the distribution in the silo. Fine grains may accumulate at the top, but the middle areas and the sides of the silo are also potential places of enrichment. This arrangement of the material in the silo may affect the distribution of material on the conveyor bands.

## 2.2.3 Problems Related to the Physical Properties of Sand

Both the chemical composition and the granulometric stability of sand are equally important, and should not be allowed for any change. The dissolution of quartz in soda ash is a determinant factor in glass melting. Each sand has its own characteristics for melting which largely depends on the grain size, Figure 8.

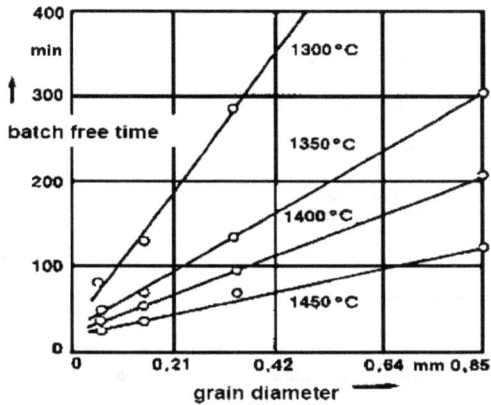

Figure 8: The batch free time is closely related to grain size and temperature. Experience shows that sand in the batch is the most effective raw material that effects batch free time.

The coarse and the fine portions should be stable, but much of the sand is preferred to be concentrated around a md-value, e.g. 0·3–0·4 mm. The fine fraction dissolves earlier than the coarse sand. There are various reasons for limiting the fine portion, one that stands out most is that it can be easily carried over to the regenerators. Due to this reason the fraction below 0·1 mm is kept to minimum value (<2–3%).

It is the personal view of the author that, it is well worth to discuss the behaviour of fine portion, in many years of experience gained from several furnaces there has not been a one to one relationship of the fine portion of the batch and that carried to the regenerators, even though the fine portions of our raw material was excessive.

It is my strong belief that the fine material is subjected to a strong sintering action due to the strong action of flames, Figure 9, thus preventing a total carry over to the regenerator. Almost an identical result has been observed in the experiment carried out for decrepitation of carbonates (limestone and dolomite). In this case, it was very clearly determined that, the fine grains were sintered and had no contribution to fly over.

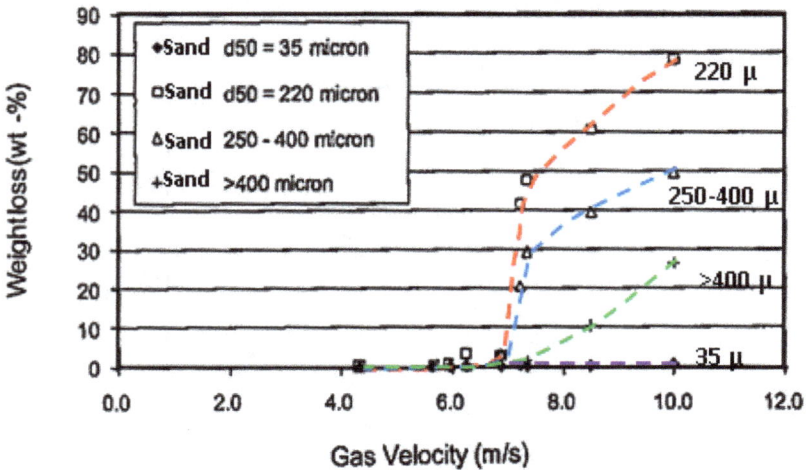

Figure 9: There is a critical velocity for the flames (7 m/s). Velocities above this value seem to aggravate the problem. Note that the fine grained particles (d50=35) do not show any mobility, they are stuck on the velocity axis. The authors experience is that the fines are rapidly sintered.

To put it in a nutshell, if all the fines were to be carried to the regenerators, glass production would not have been possible. According to classical thought, for a float size furnace the amount of fine portion that is likely to be carried over to the regenerators, amounts to several tens of tons of material every day. In practice, this is not the case. I propose that, in order to enlighten this point, some dedicated research should be made. If the results turn out to be in agreement the implications will be numerous!

One of the factors which affect carry over is the geometry of the flames in the furnace. The velocity of the flames and the angle with the batch are highly effective, there is a strong sweeping action on the batch. In practice, the velocity of flames in a float furnace are of the order of 12 m/s, in container furnaces this value is about 8–10 m/s. But as shown in Figure 9 the threshold value for flame velocity is 7 m/s, above this value there seem to be carry over potential for all sizes to some extent, except for fine particles ($d_{50}$=35 μm). This point is significant and is discussed to some length above. Note that while grain size 220 μm is critical, grain size below 35 μm has totally different behaviour [18].

Limits are also given for coarse grains in furnaces with high pull. Currently the maximum top limit for sand grain size is 0·5/0·6 mm. Above this value the dissolution becomes sluggish, and defects are unavoidable. In the 1980s the upper limit for sand was 1·0 mm, but this has been steadily reduced over the years (<0·5 mm) for a better melting rate. This is a positive approach for energy consumption. Critical sieves (0·1 mm and 0·5 mm) can be used to control the coarse and fine fractions.

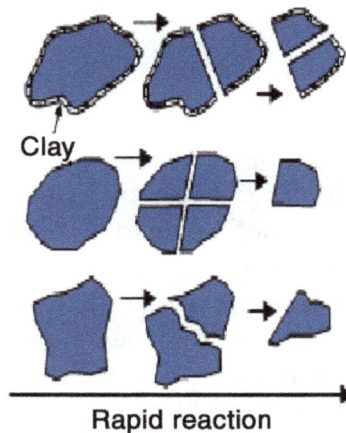

**Rapid reaction**

Figure 10: Diagrammatic display of quartz grain enveloped by a rim of clay. The reaction speed is higher with angular grains.

Melting behaviour is not only related to size, but also the shape of the grains (round, spherical or angular) is also important. Round grains are dissolved less easily than angular grains, because of the free surface available (angular grains have greater surface then equivalent round grains). This is why some sands dissolve easier than others. Some features on the sand grain surfaces may affect the rate of dissolution. For example if clay was the cementing material of sandstone during the stages of sedimentation, some clay particles may stick to the surface of quartz grains, Figure 10, the diagram clearly shows this type of occurrence. These may not be cleared off very easily. Quartz grains enveloped by clay particles may prevent soda ash reaching the grains, thus delaying melting. This occurrence is rare, but it can happen, therefore a thorough investigation of all properties must be accomplished. Scrubbing may not eliminate totally these clay particles.

## 2.2.4 Fluctuations in Limestone and Dolomite

CaO and MgO, which are essential components of glass composition, are obtained from natural raw materials of limestone and dolomite. There is a limited amount of isomorphic substitution between these minerals, in other words, there may be some dolomite (5–10%), in limestone, while limestone can be present in dolomite up to 5%. Typical compositions for these minerals are as follows:

|           | CaO  | MgO   | $Fe_2O_3$   |
|-----------|------|-------|-------------|
| Limestone | 55±1 | <1·0  | 0·04–0·06   |
| Dolomite  | 31±1 | 20–21 | 0·05–0·10   |

As mentioned before it is not only the chemical composition that counts, but granulometric properties are also important. In the 1980s the upper grain size limit for these raw materials was 2–3 mm, but currently this value is pushed down to 1·0 mm. An important point which must not be neglected for the choice of these materials is that they should not contain free quartz, which is coarser then the upper limit of sand. Otherwise there may be a potential source for silica stones, especially in float glass production. The amount of quartz may be determined by acid (HCl) dissolution. Here it is important to iterate the size and the number of quartz grains. If there is a $SiO_2$ content (>0·5%) in the chemical composition of limestone and dolomite, this test should be applied now and again, if the results are consistent the supplier should be informed.

The reason why coarse limestone and dolomite were preferred relies on the fact that coarse carbonates release their $CO_2$ content near the hot point, which would help refining and melting by stirring action. In this way, it was expected to prevent foam to go beyond the hot point. Nowadays, there are a number of different measures to apply, which can act much more efficiently.

For example during the design stage of the furnace the hot point could be defined by using mathematical modelling. If it is necessary electric boosting, bubbler or dam can be incorporated, which are significantly more effective than using coarse carbonates. Any of these applications provides a good chance to eliminate free quartz, by reducing the upper limit to less than 1·0 mm. Coarse limestone and dolomite may behave differently during melting, resulting in CaO increase in the melt, causing an inhomogeneity in glass [19].

The iron content of limestone and dolomite may show significant differences. For instance if conditions are suitable during mineralisation some iron containing carbonate ($FeCO_3$), ankerite, may also form. Iron increase may also be due to presence of clay formations in limestone and dolomite, which may give rise to the presence of $Al_2O_3$ and $SiO_2$.

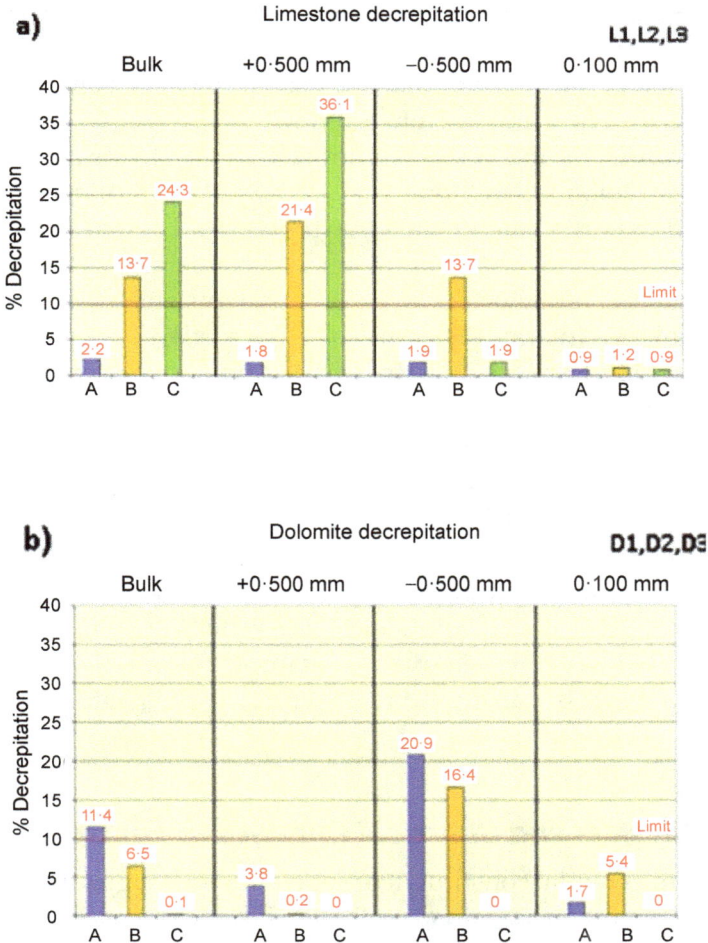

Figure 11: Decrepitation is more pronounced in limestone (a) than dolomite (b).

A very important parameter for limestone and dolomite is their behaviour of **decrepitation** [20]. Decrepitation is defined as the ability of fly over of particles during the release of $CO_2$. As shown in Figure 11, the figures for this parameter are quite remarkable; dolomite (21%) and limestone (35%). Considering the cost of raw materials, it is proposed that a limit of 10% is sufficient for decrepitation. This can be applied as the basis of selection. Very good coarse crystals of these carbonates are more susceptible to decrepitate than microcrystalline crystals. The coarser the crystals are the more decrepitation is likely to occur. It was very interesting to note during the experiments that the fraction below 100 μm did not decrepitate, but were subjected to strong sintering and agglomeration.

This means that this fraction is not carried over to the regenerators by flames. Dolomite and limestone with high decrepitation may be observed over the batch blanket as fireworks. Each of these bright sparks are particles of MgO and CaO which are transported on to the superstructure refractories like projectiles, which may trigger off corrosion, thus giving rise to rundowns into the glass melt. A potential driving mechanism for refractory corrosion and a source for knots and cords in the glass product. Any contamination in limestone and dolomite can be investigated by detailed mineralogical examination.

The determination of decrepitation may be accomplished by using the following relationship

$$\% \text{ Decr.} = \{1-(m_1/m)*100/(100-LI)\}*100$$

Where $m$=weight of crucible+sample; $m_1$=after heating and cooling, the weight of crucible+sample. The sample is first dried at 105°C, then 1 g sample is transferred to the crucible and heated.

## 2.2.5 Physical and Chemical Changes in Feldspars

$Al_2O_3$ is an essential constituent of soda–lime–silica glasses, which is obtained from natural feldspar. Turkey is a major producer of feldspar, which is also extensively used in Europe. The compositional characteristics are as follows:

$$SiO_2=70\% \qquad (60–85\%)$$
$$Al_2O_3=20\% \qquad (10–30\%)$$
$$Na_2O+K_2O=10\% \quad (4–15\%)$$

Figure 12 shows the range of fluctuation.

For those feldspars which have $Al_2O_3$>22%, it is possible that some nepheline may also be incorporated. The oxides of $Na_2O$ and $K_2O$ act as flux materials during melting. The melting points of feldspars are within the range of 1100–1200°C. Because of the presence of alkali oxides, soda ash is used less, which is a cost reduction factor. Like in all other raw materials, $Fe_2O_3$ may show significant differences and fluctuations. The most frequent Fe-bearing minerals are biotite and amphibole. K-bearing feldspars (orthoclase and microcline) may undergo some in situ alteration, where Fe is easily adsorbed by the minerals. Ti-bearing and Fe-bearing minerals (ilmenite and leucoxene) can be found in association with feldspar.

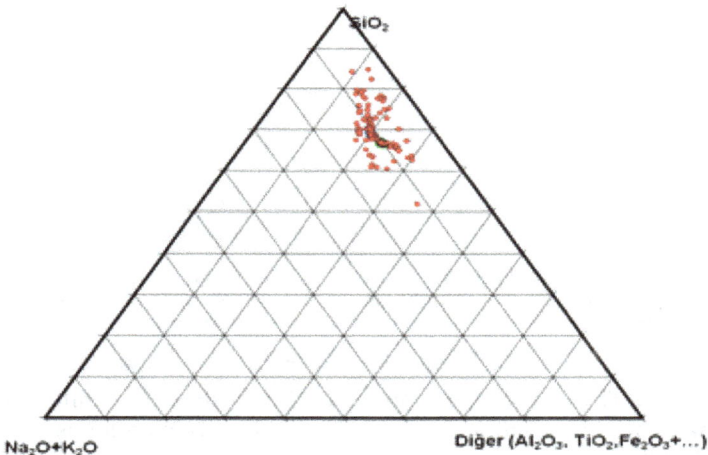

Figure 12: The scatter in chemical composition in feldspars mined in Turkey.

Apatite (Ca-bearing phosphate) is not rare in feldspars. Experience gained through the years shows that if the size of apatite ($CaPO_4$) is coarser than 700–800 µm, it may hinder the homogenisation of high viscosity aluminous material. This may lead to knot formation. If a size limit (<0·8 mm) is applied the problem can be eliminated or minimised.

This upper size limit is an indirect method of solution for coarse free quartz grains. Some years ago this limit was even greater than 1·0 mm, but currently there is a trend to reduce the grain size for better melting.

A mineral which is related with feldspars is zircon ($ZrSiO_4$), which is also a typical mineral of acidic igneous rocks, can give rise to stone defects. Nepheline syenite is more enriched with zircon; therefore an upper grain size of 0·8 mm will take care of this problem.

### 2.2.6 The Effect of Contaminants on the Quality of Glass

Glass producers of all types are on a continuous alert about the contamination of their raw materials. If they are of refractory nature they do not dissolve in the glass, this resulting in stones, or they may create a colouring problem in glass.

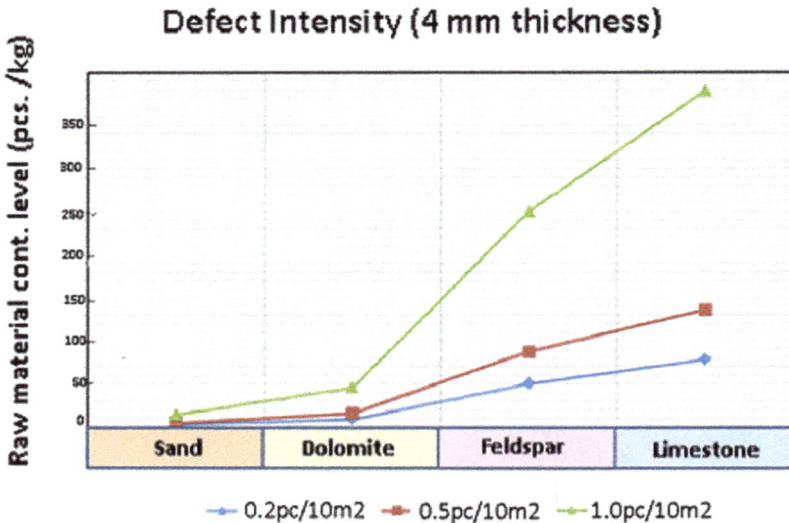

Figure 13: The quantity of contaminants contributed by different raw materials. The effect is seen clearly. Sand contamination is highly dominant in defect intensity, while dolomite and limestone are proportionally less. The calculations are made for 4 mm thick glass (the contamination level of raw materials are: sand 5 pcs/kg, dolomite 15 pcs/kg; feldspar 85 pcs/kg; limestone 135 pcs/kg).

Quite frequently the glass producers impose some protective limits about the size and quantity of contaminants. For example the following limits, for sand used in float glass production are used for chromite:

>400 μm    0 grains/ton of sand
>300 μm    40 grains/ton of sand
>200 μm    1000 grains/ton of sand

The mineral processing systems may be capable of eliminating the contaminants. For instance, Fe-bearing minerals may be recovered by *spiral magnetic separators*. In cases like this, it is vital to determine, as quickly as possible, which of the material is contaminated and to what extent. The methods to apply when there is a contamination or mixing of raw material are explained in detail (Appendix 2). It is critical at this stage, to correlate the defect intensity, in the product with the findings derived from raw materials.

Chromite grains greater than 220 μm cannot dissolve in the glass melt [21], therefore they give rise to serious quality problems. The contamination of any of the raw materials (sand, limestone, dolomite and feldspar) for a 700 t/d pull in float glass are iterated with the following example. The amount of raw materials used are: sand 490 t/d, dolomite 150 t/d, feldspar 28 t/d and limestone 18 t/d. For the same contamination level in each raw material, to what extent the defect intensity is affected; or which contamination level in the raw materials will give the same density, also taking into account the thickness of glass is iterated. A defect intensity of 0·5 grains/10 m$^2$ can be produced with the following contamination of each raw material, Figure 13.
Sand 5 grains/kg; Dolomite 15 grains /kg; Feldspar 85 grains /kg; Limestone 135 grains/kg

**The defect intensity (grains/10 m$^2$)** for different thickness are calculated for a 700 t/d pull furnace. If we have an idea about the Number of Chromite Grains entering the furnace per Day the intensities are as follows (**NCGD**);

| 4 mm glass | 6 mm glass | 8 mm glass |
|---|---|---|
| (NCGD/70 000 m$^2$)*10 | (NCGD/46 000 )*15 | (NCGD/35 000 )*20 |
| 1 m$^2$, 4 mm glass=10 kg | 1 m$^2$, 6 mm glass=15 kg | 1 m$^2$, 8 mm glass=20 kg |

## 2.2.7 The Effect of Contamination of Raw Materials in Container Glass Production

Container and tableware glass productions are highly energy and labour intensive sectors, where 40% of production costs are made up of raw materials and energy [22,23]. There are several production cost parameters, the energy consumption cost activities are indicated in the following diagram (Figure 14).

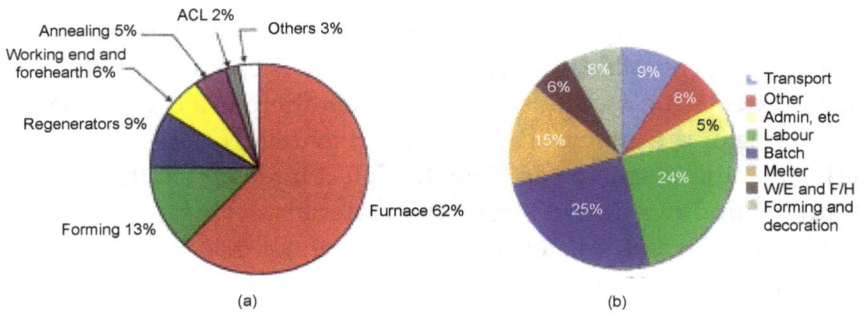

Figure 14: Energy consumption in a glass factory is made up of several activities (a) the main share is that of the furnace (62%); there are several other costs like raw materials, labour costs, transport, etc (b).

In flat glass the quality is assessed through colour, optical distortions and the number of defects present in unit area. In container glass production it is more appropriate to follow the quality with the number and type of defects in one bottle or container. If there are any refractory contaminants and if glass-ceramics find their way into raw materials or cullet there is a high risk of stones and cords.

It is evident that special care should be exercised for glass-ceramic contaminants, otherwise one can be driven mad. In case of doubt the contaminated material should not be used. If contaminated cullet is used over and over the risk gets larger, and may take a long time to clear.

If 220 μm is a critical size, like chromite, for melting the same reasoning can be applied for quality. For example, for a furnace with 230 t/d pull and 68 tons of cullet per day; the raw materials used are sand 123 t/d, limestone 19 t/d, dolomite 21 t/d.

If there is a contamination level of 1 grain/kg of sand; then there is more than 50% possibility to find one defect in each bottle. If there are 10 grains/kg of sand, then production is at a total risk. For 1 grain/kg of limestone or dolomite, if the weight of the product is not too small then quality can probably be tolerated. If there are more than 10 grains/kg of limestone or dolomite, special measures should be taken.

## 2.3 Quality Specification for Raw Materials

Raw materials like sand, dolomite, limestone and feldspar have to be subjected to some sophisticated methods of enrichment in order to satisfy the specifications set out by the glass producer [24]. The

$Fe_2O_3$ level and the granulometric properties are two parameters that are strictly controlled. Mineralogic investigations at all stages should be carried out.

Very often only chemical analyses may not be sufficient; for example, if the grain size for chromite is critical or on the coarse side, it may not reflect itself in the chemical analyses. Since coarse size grains do not dissolve, there may not be a problem with colour, but stone defects are unavoidable. In other words, if there is no problem with the chemistry, this does not mean that there will not be any stones.

### 2.3.1 Specifications for Sand

In addition to rigid specifications laid down [25–27], an all-time alert state should be exercised. Specifications are not sacred, they should be dynamic and changed and readjusted if seen necessary, by mutual agreement with the supplier. General outlines are given below, Table 9.

I. Quality>Crystal glass
II. Quality>Tableware glass
III. Quality>Flat glass and flint container

| $Fe_2O_3$, $SiO_2$, $Cr_2O_3$ levels (%) | | | Moisture (%) | Grain size |
|---|---|---|---|---|
| $SiO_2$ | $Fe_2O_3$ | $Cr_2O_3$ | 3–5 | Majority between |
| I. Quality  99·5 | 0·008 | 0·0001 | | 0·2–0·4 mm |
| II. Quality  99·0 | 0·010 | 0·0002 | | >0·4 mm, max. 2% |
| III. Quality 98·0 | 0·030 | 0·0003 | | >0·6 mm, 0% |
| | | | | <0·1 mm, max. 1% |

Table 9: The quality specifications of sand (source of $SiO_2$) for different type of glasses. Currently there are more stringent specifications as quality demands increase.

The majority of the $SiO_2$ component of sand is made up of quartz. However, in addition to $Fe_2O_3$, several other oxides namely $Al_2O_3$, $K_2O$, $Na_2O$, CaO and MgO can also be present. The presence of $Al_2O_3$ is usually related to alkali oxides, in which case the presence of feldspar and clay may be suspected as the source minerals. CaO and MgO are invariably related to limestone and dolomite. The following calculations are pertinent for further evaluation. Especially the presence of clay should be questioned thoroughly. If the clay component of sand is above a certain value, e.g. >1%, quartz–mullite defects may be inevitable in glass, which may also be highly detrimental to quality. If a suitable way is not found to eliminate the clay from sand it will be a cause of everlasting problems. In this case one might be suspicious of the clay particles encapsulating quartz grains, Figure 15.

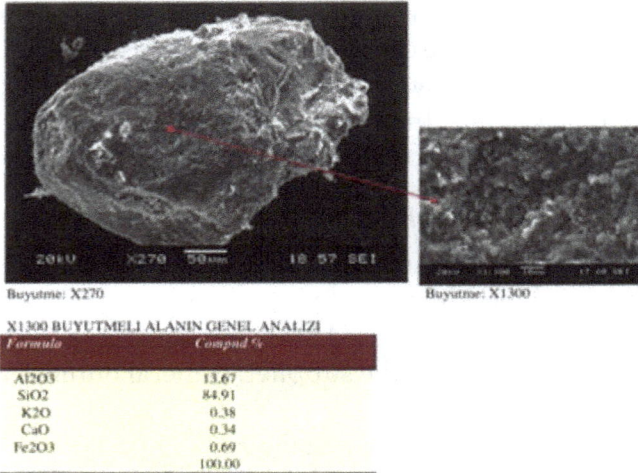

Buyutme: X270                           Buyutme: X1300

X1300 BUYUTMELI ALANIN GENEL ANALIZI

| Formula | Compnd % |
|---------|----------|
| Al2O3 | 13.67 |
| SiO2 | 84.91 |
| K2O | 0.38 |
| CaO | 0.34 |
| Fe2O3 | 0.69 |
| | 100.00 |

Figure 15: Not frequently seen, but sand grains can be encapsulated by a rim of clay, which hinders the reaction with soda ash, this is a disadvantage for glass melting.

It is possible to calculate the mineralogic components of sand by using the chemical analyses, the details of which are indicated below, Table 10. This information is very useful during decision making, for example to know the proportion of feldspar in the case of feldspathic sands.

| Component to be calculated | Calculation |
|---|---|
| Feldspar in sand (FLD) | $5 \cdot 9 * \sum R_2O$ |
| $Al_2O_3$ in feldspar (A1) | $0 \cdot 183 * FLD$ |
| $SiO_2$ in feldspar ($X_{SiO2}$) | $0 \cdot 647 * FLD$ |
| Clay in sand (C1) | $\{\sum Al_2O_3 - A1\} * 2 \cdot 53$ |
| $SiO_2$ in clay ($Y_{SiO2}$) | $0 \cdot 465 * C1$ |
| Quartz | $\sum SiO_2 - (X_{SiO2} + Y_{SiO2})$ |
| 1 CaO | $1 \cdot 784 \ CaCO_3$ calcite |
| 1 MgO | $4 \cdot 61 \ CaCO_3.MgCO_3$ |

Table 10: It is possible to calculate the proportion of some minerals (feldspar, clay, quartz, limestone and dolomite) from the chemical composition of sand.

## 2.3.2 Specifications for Limestone and Dolomite

The CaO component for glass is obtained mainly from limestone ($CaCO_3$), the rest is provided by dolomite ($CaCO_3.MgCO_3$) which is the only source for MgO. These two raw materials have a lot in common, therefore they have similar specification limits [25] and are examined under the same heading, Table 11.

    I. Quality       Tableware glass
    II. Quality      Flat glass, flint container
    III. Quality     Coloured glasses

| $Fe_2O_3$ and $Cr_2O_3$ levels in limestone/dolomite (%) | | Moisture and acid soluble, decrepitation | Grain size | |
|---|---|---|---|---|
| | $Fe_2O_3$    $Cr_2O_3$ | Quality I. and II. | Container | Flat |
| I. Quality | <0·04  <0·0004 | Acid Soluble <1% | >1 mm  <0·5% | – |
| II. Quality | <0·12 | Moisture <2% | >0·5 mm  – | <0·5% |
| III. Quality | >0·3 | Decrepitation <10% | <0·1 mm  <10% | <2·5% |

Table 11: The specifications (physical and chemical) of limestone and dolomite.

*Conversion factors for CaO and MgO are as follows:*
*% CaO in pure limestone=55% (98·1% $CaCO_3$)*
*% CaO in normal limestone=47·7% (85% $CaCO_3$), the rest (MgO, $Al_2O_3$, $SiO_2$)*
*$1CaO=1·784CaCO_3$*
*$1CaCO_3=0·560CaO$*
*Ignition loss=0·44*
*$1MgO=4·61MgCO_3.CaCO_3$*
*$1MgCO_3.CaCO_3=0·217MgO, 0·304CaO$*
*Ignition loss=0·479*

## 2.3.3 Specifications for Feldspar

Feldspar is used as source material for $Al_2O_3$, which is of low quantity (<3%) in the glass composition but an essential one for the chemical resistance of glass. However, there are other glass types where $Al_2O_3$ is a major component (aluminosilicate glasses, fibre glass, borate glasses). Albite is the major mineral in feldspars, for all the other raw materials options the $Fe_2O_3$ level is an important selection criteria, for general approach an index which combines $Al_2O_3$ and $Fe_2O_3$ is developed ($Fe_2O_3$ index=$Fe_2O_3/Al_2O_3$), but quite frequently K-bearing feldspars (orthoclase and microcline) can be in association with one another. The majority of feldspar used in Europe is produced in Turkey [12]. Like in

*I. Quality (colourless glass), $Fe_2O_3$ index=0·006*
*II. Quality (flat and container), $Fe_2O_3$ index=0·012*
*III. Quality (coloured glasses), $Fe_2O_3$ index >0·012*
*Grain size <0·8 mm*
*Ignition loss <1%*
*Fluctuation between parties $R_2O$=±0·5%*
*$Al_2O_3$=±0·25%*


***Conversion factors:***
*% Feldspar (FLD)=5·9\*$\sum R_2O$*
*% $Al_2O_3$ in feldspar=0·183\*FLD*
*% $SiO_2$ in feldspar=0·647\*FLD*

# CHAPTER 3

## 3.1 Some Problems Encountered in Glass Production and Glass Quality from Batch to Warehouse

The sequence of events, in the good and negative sense, start as the batch is charged to the furnace, which continues until and includes the forming operations, and once the product is in the warehouse different developments occur. This sentence should not be taken as there are problems all the time and round the clock. A better expression would probably be that all production types have their own problems. When efficiencies are high disturbances are at low level, in which case the morale and motivation of the personnel is rather high. In cases of unforeseen disturbances, the limits on quality are at a low level; here an emergency state is declared for the personnel involved.

The author had the chance to be involved in numerous case studies which have provided a wealth of experience over the years. A valuable advantage was to have had a chance to access several type of furnaces (float, container, tableware, E-glass, crystal and borosilicate), and very close cooperation with the mining personnel. This experience would not have been consolidated if strong team ethis was not exercised with the furnace operators and R&D engineers, etc. An attempt will be made here to explain several areas of problematic nature.

## 3.2 Some Problems in the Furnace Bunker

If the raw materials are delivered to the glass producer by the supplier abiding by all the specifications, then all the responsibility belongs to the producer after the raw materials are charged into the furnace. Batch preparation is the first stage in production. Raw materials are weighed in automatic systems and are transferred to the mixer. To avoid any segregation due to varying grain sizes of the raw materials some water or water vapour is introduced into the mixer. The mixing action is realised over a suitable time and then the whole mixture is charged to the furnace bunker. The capacities of mixers vary between 1–5 tons. The batch during transfer from mixer to the furnace bunker may lose some moisture depending on the distance travelled. Fine grains may segregate to some extent (7). Some parts of the batch may accumulate around the inner surface of the bunker, or even they may form a compacted layer near the bottom. The thickness of these accumulations may be 8–10 cm thick, which may cause a narrowing for batch flow. They have typically a banded structure, light and dark coloured. To prevent or minimize this type of formation the bunker is hammered by automatic systems.

Investigations carried out on the material described above has yielded 10–30% clay and high alkali level (35–55%). Fine sand is also enriched in the banded structure. This banded material is not seen as a problem for melting, as experience proved.

The main problem for the bunker is if the temperature of the batch falls below 40°C. Soda ash ($Na_2CO_3$) may easily adsorb the free moisture below 35·4°C and form a new crystalline phase, $Na_2CO_3.10H_2O$. Under these circumstances the batch becomes quite dry, which causes dusting and prevents well defined log formation. In some cases in order to prevent this problem a heating facility may be applied around the bunker. It is again worth noting that the batch moisture should be 3·5–4·0% to ease the log formation.

## 3.3 Batch Homogeneity

The most important feature for the batch is that it should be homogenous to the best level possible. Each $cm^3$ of the batch should be of identical character. The following parameters should be optimised for the particular mixer; discharge times, rotation time per minute and the amount of moisture. Excessive mixing may cause energy consumption and desegregation of the batch. On the other hand, insufficient mixing is likely to result in inhomogeneous batch.

An appropriate method to determine batch homogeneity can be accomplished by taking equal amounts of batch (5–10 samples) at fixed intervals and repeat the same for different mixer discharges (at least five times and same number of samples). All the samples can be analysed by XRF. The samples are prepared by pressing. The measurements bellow can be accomplished.

- To start the experiment it is strongly recommended to prepare a synthetic batch (without cullet) and repeat measurements at least 3–5 times in order to determine the oxide variations. It should be possible to define upper and lower control limits for oxide variations. These are to be used as reference values for real batch measurements.
- Use 3–4 mixing times, do the same sampling; in this way a suitable time can be determined for mixing, given the best homogeneity. Number of rotations per minute for mixer is also another parameter, this also can be investigated if needed. It is preferable to do the experiments at a fixed rotation.
- The fluctuation of oxides between different batches gives information about the homogeneity of batches that enter the furnace.
- If there is no XRF facility, then acid solubles (carbonates) and insolubles (sand and feldspar) can be used, the same procedure should be applied to synthetic batch.
- It is important to point out that all the above studies are unique for the mixer studied, it is evident that all these studies must be carried out for each mixer.

After mixing the batch is discharged on a conveyor band, on top of which cullet is emptied for the same period. It is established by research that cullet dust is highly harmful to refractories. The spread of batch blanket and its shape in the furnace has a lot to do with the way it is charged by the charger. In a float furnace the batch will be subjected to a temperature of 1450°C between ports 0 to 1. This temperature has a direct effect on the level of $SO_3$ in glass. The higher the temperature the lower the $SO_3$.

## 3.4 The Behaviour of Batch in the Furnace

As the batch enters the furnace it encounters a high temperature environment, where melting or dissolution of raw materials commence. If the grain size of raw materials is suitable and if there are ideal conditions for melting, then thermochemical reactions start. Some of the materials (sand and feldspars) undergo dissolution while $Na_2CO_3$ and $K_2CO_3$ start to melt, dolomite, limestone and $Na_2SO_4$ start to decompose at later stage. After the release of $CO_2$ from carbonates the other reaction products CaO and MgO start to dissolve like $SiO_2$. **Decrepitation of carbonates** takes place at this stage, where pyrotechnics may be observed. The first reaction to take place in the batch is the vaporisation of free moisture at about 105°C, Figure 16 followed by soda ash, Na-sulphate, dolomite and limestone.

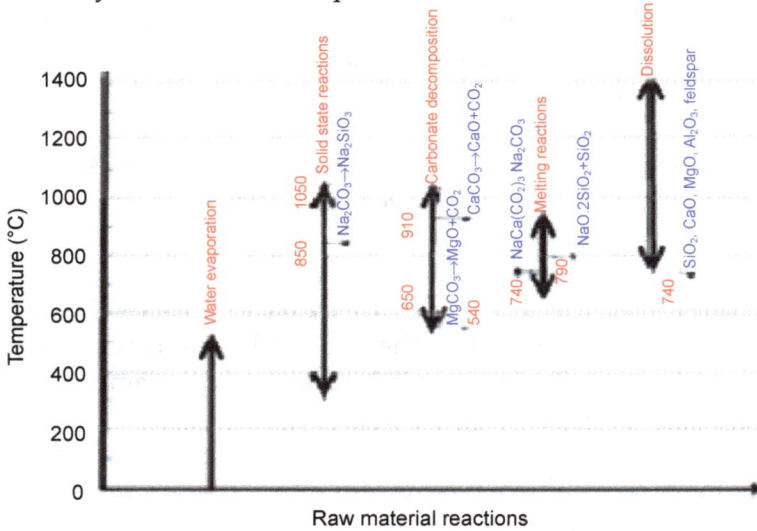

Figure 16: Melting and decomposition temperatures of glass raw material.

Once the dissolution of quartz is accomplished and high viscosity glass has formed, then the homogenisation of glass melt starts, Figure 17. Homogenisation of the glass melt can be considered complete after the release of bubbles (fining and refining) is almost finished. The quality of glass melt depends on whether it has completed all stages satisfactorily. Irrespective of the glass type (float, container and tableware) the aim is to get a glass quality which has the minimum number of defects per unit weight. The defects which may deteriorate the glass are discrete defects like stones and bubbles and vitreous defect types, e.g. cords and knots [28]. If there is a satisfactory level of homogenisation then forming operations can start.

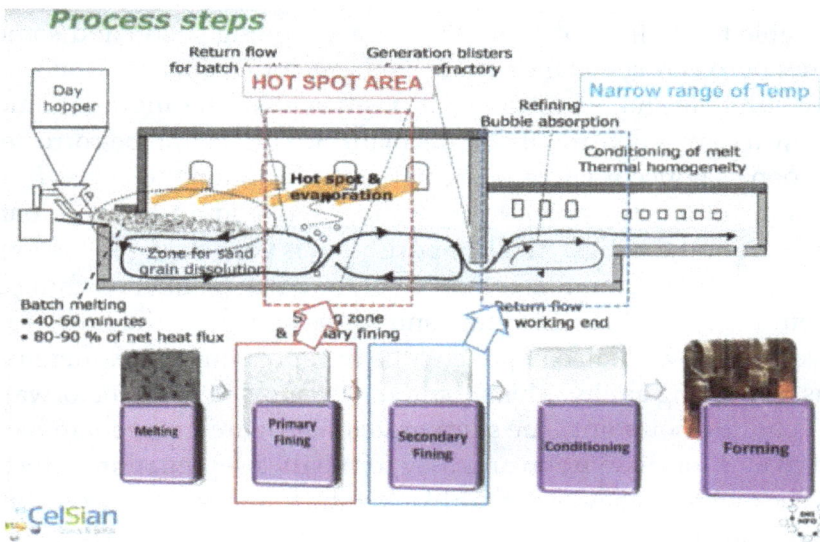

Figure 17: After the quartz grains are dissolved in the first half of the furnace then homogenisation (including primary and secondary fining) starts, which is followed by conditioning. There is no clear cut separation between these stages [infodesk@celsian.nl].

The batch is charged on molten glass, where it is under the effect of flames on the top with a forward movement, while the bottom of blanket is under the action of backwards moving glass currents. Batch reactions continue both on the top and bottom of the blanket. The energy supplied by convection currents to the glass which is the main force driving it forward is affected by two energy sources, flames on top and the convection currents below. The following criteria have to be satisfied for batch to transform into glass:

- The batch has to be mechanically pushed forward, accomplished by batch charger,
- For efficient energy transfer to the blanket surface, at best, it must be undulated, which provides a maximum surface,
- If disintegrated small parts of the blanket may still move forward; the glass currents should take care of these pieces so that they are returned back towards the doghouse, with the high possibility that they will mingle with new incoming batch to complete all the reactions missed out.

When the batch blanket is in the 1400–1450°C temperature zone, melting on the surface and liquefaction can easily be observed. It is possible to see flow off from the surface, boiling action and sometimes pyrotechnics on the surface. Soda and other alkali components (Na-sulphate and Na-nitrate) will easily melt. If the angle of flames is low it may touch and fly off the blanket. This should be corrected as soon as noticed. There is a remarkable flow of molten glass from the hills of batch blanket to the valleys. This liquid forms a foam layer around the logs. Alkali vaporisation is very strong in the long run, this foam accumulates in front of the batch blanket reaching to about 1 m of width in the centre and a little more near the sides. This layer has a low alkali component due to vaporisation, and therefore has no melting ability. This should not be allowed to travel forward beyond the hot point. The silica grains which are in the foam have no chance for dissolution and they form **silica segregation defects**.

The melting behaviour of the blanket shows differences depending on the side of flames. The area under the flames has less logs, where energy transfer is more effective. Fining of the glass melt changes and some kind of inhomogeneity develops. This situation changes during the inversion of flames. In end-fired furnaces this effect is more pronounced. This type development gains a continuous manner.

## 3.5 Furnace Observations

The main objective of furnace operators must be to observe the melting process in the furnaces they are in charge of, irrespective of the size, small or big. The furnace is a determinant factor, with all the activities around it. There is no doubt that the furnace is the heart of the glass production business. Everybody in the campus whether technical or not has an interest in what goes on in the hot area. Of course, all the links of the chain are undoubtedly important, but the furnace is a little bit more important. There cannot be any tolerance for any mishap in any activity, the wheel must go on turning without any interruption. For example, if glass of high quality, is not packed satisfactorily, or not stocked under good conditions in the warehouse and finally if customer satisfaction is not attained in this transaction, the whole activity is a failure. The morale and motivation of all people is very important for good results. In other words, if everybody does their own best, there should be no problem.

The responsibility of people working around the furnace must be higher for the good of the establishment. If at the end of the day everything goes according to schedule without deviations, however tired they are during the day, they may even feel happy. On the other hand, if there are problems with the furnace, everybody has their share of sweating. No matter what may have happened at the end, success will come, it is at this time when everybody feels happy, forgetting their tiredness. In the case of any delicate change or disturbance, the furnace must be carefully observed before, during and after for satisfactory evaluation. Observations made by different people may differ and thus some useful information can be obtained. Two plus two may not be four for different observers in furnace observations. There are many ways to describe flames: long, short, bright, smoky, lazy, disrupted, angled down or up, etc. What is more the observation ability for everybody varies remarkably. The observer must be able to view the forest to get a general view, and yet at the same time, when necessary, he should be able to view individual trees. Under all circumstances the observation must furnish some new information [29]. Furnace operators, because it is their own job, must everyday make regular and on-and-off observations. The points which come out from observations may necessitate some precautionary measures, which are in the long run prolonging the lifetime of the furnace. It is apt to say at this point that some are more talented and capable in making furnace observation. From these experiences some concrete points may stand out, which must be noted in the next campaign or repair of the furnace. It is important to note at this stage that enough observation holes with suitable physical condition should be put in the furnace, whatever the size, so that detailed observations can be made. The security and safety of the furnace operators should be number one issue in the list of priorities.

Traditionally furnace observations are made with dark blue or dark green watching glasses. Although they seem to serve the purpose, for detailed observations these are insufficient. In actual fact, it is not only the glasses which are not good enough but weakening of vision of the eyes of personnel is more of a problem. The resolution of normal eyes is 25 μm at 75 cm, and 1 mm at 450 cm. In normal daily life, one quite often needs glasses to watch TV, or in a conference, when necessary, far sighted glasses must be used to overcome the problem. Similarly, when making furnace observations, the author strongly advises the use of facilitating tools such as a monocular or binoculars (a blue glass must be inserted in the ocular). It takes a short time to get used to them, and they will be found to be extremely beneficial.

Several other facilities (periscopes, water cooled cameras, TV monitors) are available to furnace operators. Currently, the endoscope is the ultimate instrument, this is capable of observing critical places, which are otherwise inaccessible with blue watching glass; these instruments have the ability to zoom into these points for detail. Detailed observations provide vital information, so that more satisfactory decisions can be made for repairs.

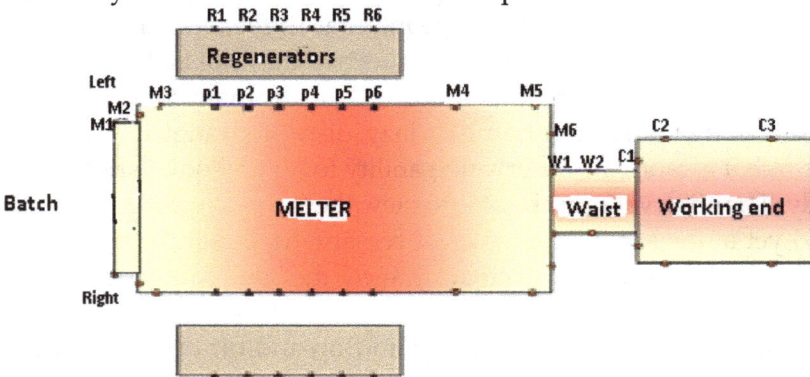

Figure 18: Observations of the furnace is so important that it must become a 'routine practice' for those personnel who are operators and for those who are superintendents. The points from which observations can be made are indicated as: **M1-M6, W1-W2, C1-C3, R1-R6, P1-P6.** A summary of the observations which can made are listed below:

*M1-M6 (Melter): Batch charging, D/H coolers, log height and shape, glass level erosion, tuck stones, logs scraping the side blocks, logs beyond the hot point, sparks over the batch (decrepitation), D/H wall details, superstructure up to P1, foam, details of bridge wall, drippings from tuck stones, end of melter, crown state, accumulations in front of waist coolers, bubbler features, batch–hotpoint–foam relationship, flame angles and brightness impingement on superstructure, mirror etc.*

*W1-W2 (Waist): Waist structure, coolers and accumulations, stirrers and rotation features, cover blocks etc.*

*C1-C3 (Conditioning): Surface quality of glass, any inversions, crown and its state dry or wet, canal features etc.*

*P1-P6 (Ports): Port crowns, drippings, port bottom, displacement of refractory blocks, any dark points at the opposite side below tuck stones, etc.*

*R1-R6 (Regenerators): Regenerator chequer bricks, accumulations and congestions, crown and superstructure refractory state, port neck and bottom, drippings on port arches, watch tip of flames dark bright or entering regenerator, etc.*

The approach applied for furnace observation in end-fired and cross-fired furnaces is essentially similar, but some details may be required where necessary. The basic target in both furnaces is to get the best possible condition for glass melting. The differences in forming processes may necessitate some special observations, for example in the forehearths of container and tableware furnaces observations serving the purpose must be made. It is evident that operators are well furnished with a wealth of experience, but for trouble shooting cases certain deviations must be made. To serve such purposes the following Figures 19 and 20 will provide some guidance for where the observations should be made and outline what needs to be observed, Figure 19. If firing is on the left hand side observations can be made at points M1, C2, M6 and R. Identical observations should be made from the right hand side, just after the inversions of flames. If parameters like flames or dark points where there is potential air ingress, rundowns at glass level, are considered to be important then one can get involved in detail, but be careful not to neglect other points also, at least have an overview. The most satisfying observations (except flames) can be accomplished during the inversion time, which is one minute maximum. In this interval one can observe both ports in end-fired furnaces, but in cross-fired furnaces only two adjacent ports can be observed. Therefore, in a float furnace at least three inversion times should be used. Voice recorders can be used for taking notes during this activity.

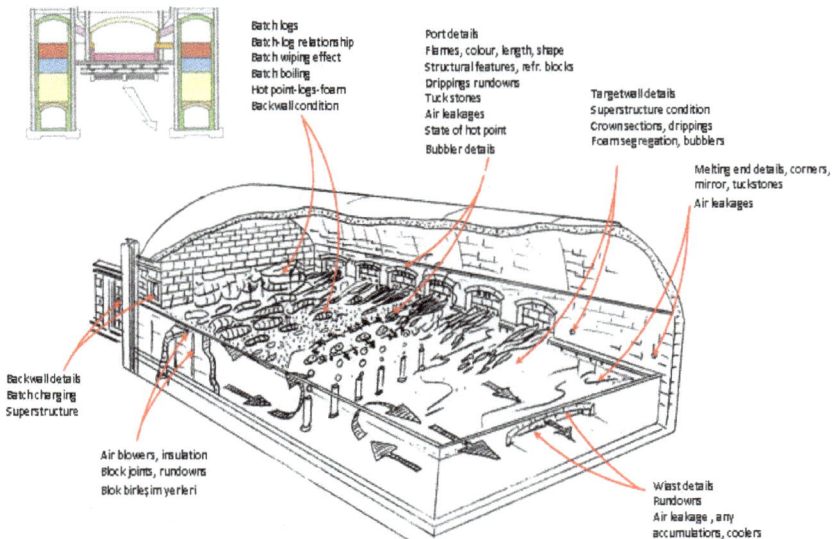

Figure 19: A section of a melter shows what you must note during observation.

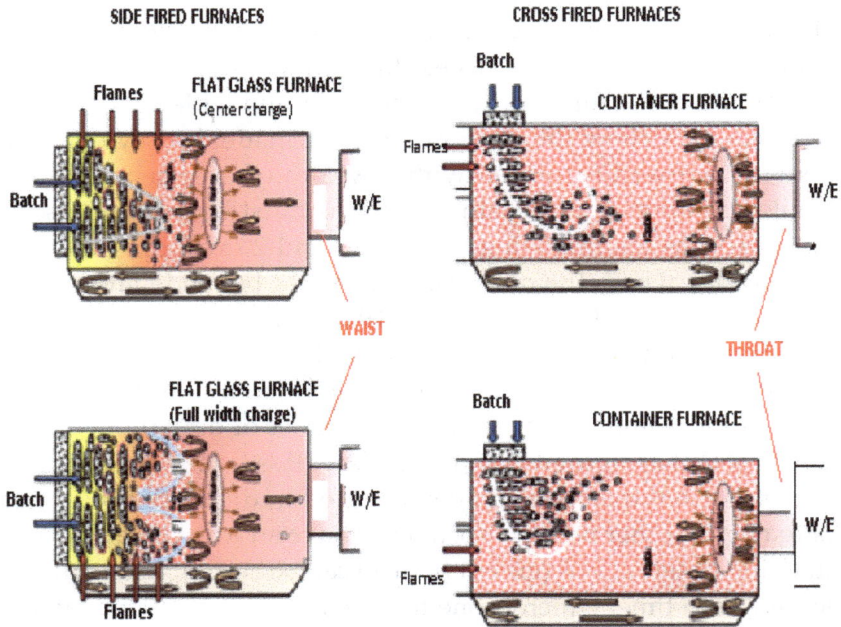

Figure 20: Side-fired float glass melter (left 2 figures) and end-fired container or tableware glass furnaces (right 2 figures). Batch charging is central on top left, and full-width left bottom. The spreading of logs is different, hot point at the last 1/3.

In container furnaces batch is charged from one side (maybe both sides). The distribution of batch coverage changes as the flames change sides. The hot point may be near the target wall, the furnace may seem to be covered with foam, with a very small mirror in front of the throat.

# CHAPTER 4

## 4.1 The Role of Moisture and Physicochemical Changes

The addition of moisture up to a certain level has become a daily practice over many decades, and it also provides some technologic advantage, which have plausible explanations. In this sense moisture has been accepted as an integral component of the batch in soda–lime–silica glasses. It is not only the amount of moisture added, but several other parameters that have a direct effect on the behaviour of batch:

- The temperature of raw material,
- The granulometric properties,
- The temperature of water used,
- The stability and flow rate of water,
- Nature of water used (hard or soft),
- The nature of water vapour used,
- The addition method of cullet to the batch,
- The decrease of temperature and water content during the transport of batch to the furnace.

Raw materials delivered by the supplier are discharged in specially prepared places (silos or covered areas in the stock hole). The temperature of raw materials is that of the stock hole. As the temperature of raw materials is low in the winter, the water used must be of suitable temperature:

Temperature of water used in the summer 60–70°C
Temperature of water used in the winter >70°C

Some special precautions must be taken for soda ash, because it is water soluble, starting from the stock area.

After mixing of the batch the temperature must not fall below 36°C. Otherwise the free water will be adsorbed by soda ($Na_2CO_3$) to convert into crystalline forms monohydrate ($Na_2CO_3.H_2O$), heptahydrate ($Na_2CO_3.7H_2O$) and decahydrate ($Na_2CO_3.10H_2O$), which means that free water is consumed and therefore the batch will become dry. To prevent this development a minimum temperature of 43–46°C must be maintained [31].

In addition to the above changes sulphate and nitrates may dissolve to a certain extent. Soda also dissolves to some degree giving rise to the formation of metasilisic acid ($H_2SiO_3$) and orthosilisic acid ($H_4SiO_4$). At temperatures above 50°C some caking may happen in the batch due these acids.

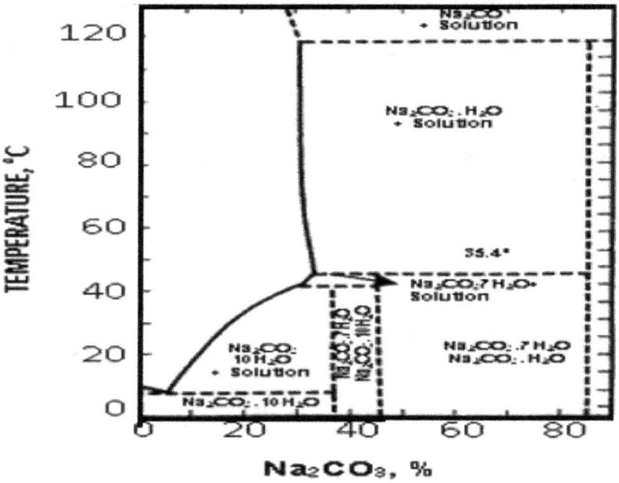

Figure 21: The behaviour of $Na_2CO_3$ and its hydrated phases is clearly related with temperature. The formation of hydrated species must be prevented by keeping the temperature above 35°C. This temperature is critical for batch at any stage. Otherwise water will be adsorbed by soda to form very dry batch.

As seen from phase diagram, Figure 21 some developments may happen in the batch, Table 12. Under all circumstances the temperature of the batch and that of soda should not be allowed to fall below 35·4°C. Comprehensive information dedicated to changes in soda ash is given in Table 12, [32].

As explained above the degree of crystal development depends on several factors:

- The temperature of batch at the end of mixing,
- Homogenisation of batch itself,
- The method of transfer of batch to the furnace.

| | Na$_2$CO$_3$ Pure soda | Na$_2$CO$_3$.H$_2$O Monohydrate | Na$_2$CO$_3$.7H$_2$O Heptahydrate | Na$_2$CO$_3$.10H$_2$O Decahydrate |
|---|---|---|---|---|
| %Components Na$_2$CO$_3$ H$_2$O | 100 - | 85·48 14·52 | 45·7 54·3 | 37·06 62·94 |
| Thermo-chemical behaviour | Na$_2$CO$_3$> Na$_2$O+CO$_2$ at 400°C Max. solubility in water, at 35·4°C. 49·7 parts dissolve in 100 parts of water. The solution is 33·22% Na$_2$CO$_3$ | Below 35°C it adsorbs water and the batch becomes dry, flowability increases. At 109°C loses its water. | At 35·4°C loses 6 mols of water and converts to monohydrate. Metastable between 32–35·4°C soda is in heptahydrate form the batch is dry. | Stable between −2·1–32°C It melts at 35·4°C and converts to monohydrate 0 mol water the batch is dry. |

Table 12: The thermochemical behaviour of Na$_2$CO$_3$ and its hydrated species are explained in detail.

The following points can be deducted from research carried out on batch [32].

• After mixing of the batch the best range for moisture is 3·8–4·0%
• Severe dusting may happen at points of discharge from the conveyor bands, if the moisture content is less than 3·5%
• Some difficulties may arise in transport if moisture is more than 4·5%. If the batch has to wait 4–5 h some caking may happen, which may present melting difficulties
• 42 litres of water/ton of batch gives a moisture level of 3·8–4·0% moisture. During the transport of mixed batch to the furnace some deterioration may develop in batch homogeneity, a decrease in moisture and also a decrease in the temperature of batch can occur. These deviations are related to the following factors:
• The distance the mixed batch has to travel. The temperature of the place where the batch is transferred through or the very last point it is transferred to may be low or with seasonal variations,
• If there are many transfer points between the mixer and the furnace.

The temperature of the batch may drop from 50°C to 30°C, while in the summer this could be 60°C to 40°C. Under these conditions the batch moisture is reduced by 0·2–0·5% in the winter, in the summer moisture is reduced by 0·5–1·0%.

When cullet is added to the batch its distribution may be irregular. The moisture level in the mixer at the beginning is about 4·0–4·2%, at the doghouse, these figures may fall to 3·4–3·6%.

If the temperature of the batch falls below 20°C some deterioration may occur in the homogeneity, due to recrystallisation of soda to hydrated forms, dusting and flowability may increase. In this way free moisture is consumed. Heptahydrate and decahydrate in the heated batch gives an exothermic reaction, therefore there is no extra energy used.

If the overall properties are satisfactory then log formation and batch distribution in the furnace will be as expected. If the moisture level is less than 3% log formation will not be good, but rather a flat blanket will form.

As mentioned before undulated batch has larger surface area, therefore energy transfer to the batch will be much better, which will accelerate the batch reactions. This will only be possible if the batch temperature is above 36°C, under these conditions segregation will be prevented and log formation is more satisfactory.

Once the batch logs are on the top of the molten glass, they will be under the control of glass currents, they will move in the same direction; forward, backward or sideways. This situation can be observed easily.

Where the logs get thinner, this is usual at the edges, there is an increased energy transfer to the batch which speeds up batch reactions, and also melting rate. Energy transfer to the batch from the molten glass below is 3–4 times more effective than the energy transferred by flames. If the batch homogeneity is well enough in all aspects the melting efficiency will be much better, which in turn means better quality glass.

It is widely accepted reality that excessive moisture means more energy for vaporisation. The findings of research carried out in USA, sponsored by the Department of Energy has shown that every 0·5% moisture in the batch requires an additional 1% energy consumption [33].

On the other hand, experience shows that the moisture in the batch improves homogeneity and once the batch is in the furnace it upgrades batch reactions. However excessive moisture may give rise to difficulties in flowability.

The presence of moisture in the batch may increase the angle of heaps, which leads to better batch feeding, resulting in an undulated blanket and more vigorous physicochemical reactions.

In the simplest way, moisture in the batch can be considered to behave as a binder, holding together the batch. If everything develops well then optimum energy will be used. Segregation is under control with suitable moisture level.

If the granulometric properties of raw materials are more or less compatible (the best preference) then no complications can occur at the stage of batch preparation or batch transport. The batch blanket in the furnace is under the influence of flames.

If the geometry is not suitable flames can sweep off the batch and cause some carryover into the regenerator. This development is also valid for superstructure refractories. The corrosion rate is highly increased therefore glass quality problems become unavoidable. This effect is more pronounced for batches with low moisture level.

To investigate the extent of carryover to the regenerators, the traditional approach is a paddle test. A water cooled, rectangular shape (3×10×20 cm) of Cr–Ni steel is prepared. This is inserted under the regenerators or over the chequers, inserted from the observation hole, Figure 22.

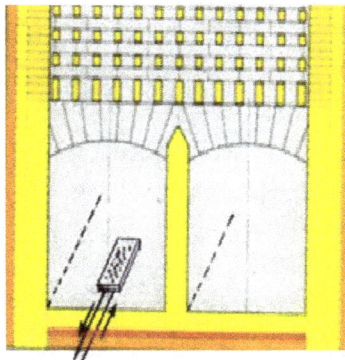

Figure 22: Carry-over of glass raw materials is measured by the paddle test, it can be put over the chequers or at the bottom of regenerators. It is not an absolute measure but provides a relative indication of what has been carried to the regenerators.

In cross-fired furnaces several ports should be investigated over a 24 hour sampling period. More than 90% of the collected sample is Na-sulphate. The water insolubles should be thoroughly investigated by mineralogic methods, detailed with electron probe if available.

The assumption made for calculation is that the area of the paddle represents a very small portion of the chequer area. It may be informative to repeat the test with different moisture levels or if there is a major change in the raw materials.

# CHAPTER 5

## 5.1 Glass Defects

As mentioned before glass production of any type is susceptible to several problems. In other words, it is not devoid of problems, with small or big impact. It is essential to keep everything under control and within tolerance. For instance, if the temperatures are low and without good distribution, then it will not be surprising to encounter batch related problems of insufficient melting or dissolution. On the other hand, if the temperatures are too high, then there is excessive vapour formation in the furnace, which accelerates refractory corrosion. This leads to the formation of discrete glass defects (stones, knots, cords) [34]. Depending on the temperature regime and the glass currents some localities in the melting end and the working end may become stagnant. Potential places are the bottom, the corners and the vicinities of waist cooler and stirrers, where glass may undergo devitrification. If the temperature of molten glass falls unexpectedly, similar developments may occur at the bottom. Devitrified glass is a knife edge situation. If due to increased temperature melting of devitrified glass develops, then there is an increase in seed level, the gas content of which is typically high $CO_2$. In this glass some crystals of devitrification, all candidates for a glass defect, may be found.

Defects found in glass can be examined by employing several analytical methods [35–37]. Defects which disturb the glass quality may originate from several sources. Forming defects form a reasonable proportion of all defects but deserve the right to be investigated separately and with an expert handling.

Defects related to the melting process may be traced back to the batch, melter, working end and forehearths. According to the source defects can be investigated under three headings: batch related defects, refractory related defects and devitrification defects, Table 13.

| STONE DEFECTS ACCORDING TO SOURCE | | |
|---|---|---|
| **Batch defects** | **Refractory defects** | **Devitrification** |
| • Related to insufficient melting<br>• Related with source and processing<br>• Contaminations | • Glass contact refractory<br>• Superstructure refractory<br>• Regenerator refractory | • Cooling of glass<br>• Remelting of devitrification |

Table 13: A simple classification of stone defects encountered in glass production. All possible source types are covered.

## 5.2 Batch Related Glass Defects

Batch related defects can occur in all types of furnaces, irrespective of the size. These are frequently related to some problems associated with the geological setting of the quarry (quality variation), contamination mixed in the raw materials in several ways, abrupt changes in grain size (coarsening) or improper melting conditions (insufficient temperatures, deviation from batch recipe) may all lead to batch defects.

'Batch' is defined as the mixture of raw materials in known proportions as prescribed in the recipe. After weighing they are filled in the mixer for homogenisation. If everything goes well, this should ideally imply that the materials are free of any problem, they are weighed accurately followed by mixing and finally transported to the furnace. If there are changes in the specification and unpredicted changes in the weighed material then the melting process will be adversely affected.

Some of the raw materials (soda, potash, Na-sulphate, nitrates and feldspar) act as fluxes in the melting process. If soda ash is excessive or insufficient the dissolution behaviour is directly affected. The most frequent batch defect is one of several forms (polymorphs) of $SiO_2$ (quartz, trydimite and crystobalite). The most critical activity in glass melting is the dissolution (not melting) of sand (quartz), which may be due to several causes; insufficient flux materials, low temperatures, coarsening of raw materials, agglomeration, irregularities in batch distribution. Most frequently conversion to trydimite develops with associated minor amounts of crystobalite. Mineralogic investigation is a powerful tool for unravelling the puzzle. We must be aware of the fact that, a good analysis forms a good part of the picture. It should be complemented with furnace operation parameters, based on an understanding of 'task force'. At first sight, if the problem is described as insufficient dissolution, the most useful remedy would be to increase soda ash and/or temperatures. This is what all glass producers apply frequently. In this case the seed level is also positively affected. It is typical to see batch stones and seeds together in case of low soda ash and temperature. Once glass quality is satisfactory after necessary precautions applied; then soda and temperature levels can be trimmed off in careful steps.

When evaluating silica ($SiO_2$) defects one point should never be neglected, that is the crowns of most furnaces and the superstructure in the flameless zone are all made of silica refractories. Therefore, any defect originating from these sections is also $SiO_2$, which may have a very similar appearance to batch related silica defects. The majority of mineral phases in refractory defects is crystobalite, the rest being trydimite and glass phase. It is the reverse in batch defects, but with many features in common. At this point one must be very careful, confusion is not unusual. It is due to this reason, the investigation of silica defects is a tedious work [38,39].

However, it is possible to unravel the problem with an electron probe instrument. If the glassy phase of these defects is analysed, it will be seen that the batch related $SiO_2$ defect has $Al_2O_3$ and $MgO$ in its composition, very much like the composition of glass; whereas these oxides are absent in the glassy phase of silica refractory. This is a clear-cut basis for differentiation. It is worth reminding that $CaO$ is not a good indicator since it is present in silica refractory as well.

## 5.3 The Formation of Foam in Glass Melting

The logs are under the influence of flames all the time; the molten material accumulates between the low levels of blanket which in turn take the form of foam due to excessive alkali vaporisation [40,41]. There is a preferential accumulation of foam at the end of the batch blanket, the width in the centre may be up to 1 m and a little more at the sides. Foam can accommodate many silica defects in it. Foam develops where alkali vaporisation is maximum, Figure 23, the amount and the distribution has much to do with temperatures and glass current activity.

1)  (Batch melting)              2) (Refining)
    Primary foam                 Secondary foam

1)  (Batch melting)              2) (Refining)
    Primary foam                 Secondary foam

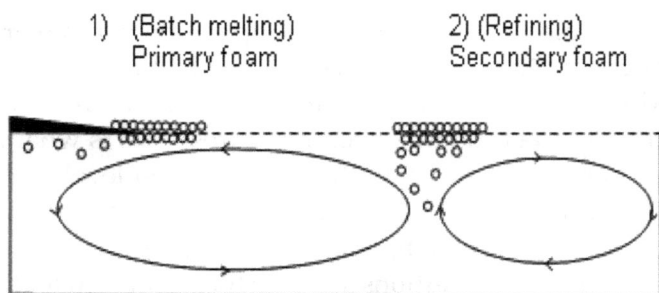

Figure 23: The formation of foam is encouraged in two parts:
end of blanket and hot point.

If the forward current is strong enough the foam can split up and may escape to the end of the melter. In float and container furnaces foam is typical. If the chemistry of batch is not suitable foam can occur. Especially the behaviour of sulphate and carbon are very critical in this sense. In float glass production the use of carbon has become traditional along with refining agent Na-sulphate. The reaction of sulphate with reducing carbon is very sensitive. There are positive contributions for glass quality. If the reaction conditions between these two components deviates during the melting process, it may be difficult to put things in order for some time. Sulphate wets the quartz grains and speeds up the dissolution process.

The optimum Na-sulphate level in float glass is 0·7–1·0 kg/100 kg glass. Carbon has a positive effect on the high pull and high quality glass. If one of these components is excessive or insufficient in float glass the impact of these minor ingredients on melting may be unexpectedly severe.

Foam must be present in the process, if there is no foam in the furnace two things may be inferred: one is that sulphate is insufficient or carbon is excessive. Anthracite or charcoal is used as a source for carbon. The burning of carbon directly affects the reaction behaviour of sulphate in two ways:
- 35% anthracite burns before 720°C, the remaining 65% reacts with sulphate,
- 65% of charcoal burns before 820°C, the remaining 35% reacts with sulphate.

Reaction at low temperatures consumes carbon and leaves behind an insufficient amount for the reaction with sulphate. As sulphate is the main refining agent there will be an obvious deficiency. At excessive sulphate consumption then the quartz grains will not be sufficiently wetted, curtailing dissolution. If there is not enough alkali then trydimite and crystobalite devitrification can occur. If sulphate is in excess then the quartz grains will not come together thus preventing devitrification; in this way the dissolution of silica grains will be accelerated. In this case refining is also positively affected. The remaining sulphate decomposes at 1440°C to $Na_2O$ and $SO_2$; this reaction takes place near the end of the blanket where the dissolution of the remaining sand grains is completed. The released gases $SO_2/SO_3/O_2$ stir up the glass speeding dissolution up. At the same time the fining of glass is also boosted. If there is foam which is made up of trydimite and crystobalite then defects reaching the pulled glass are unavoidable. To minimize the formation of foam and not allow it to move forward towards the end of the melter, the backward glass current must be strengthened. In this case it is advisable to increase the temperature difference between the hot point and the batch entrance at port 0–1 by reducing the doghouse temperature. The density of glass here is lower than that elsewhere, therefore it swims on the surface and easily finds its way to the pulled glass.

Borosilicate glass is very sensitive to foam (**scum**) formation. The major component is crystobalite due to excessively high temperatures. Vaporisation is severe. Crystobalite is of higher density than glass, therefore it mixes with glass easier.

## 5.4 Raw Material Related Defects

The properties of some raw materials may lead to problems during melting even if their physical and chemical properties are suitable. For instance clay mineral association with sand is a good example for this. Clay can be found in sands of sedimentary origin, which is very powerful cementing material in sandstones. A small fraction of clay may always be present in sands. One thing is critical, if clay particles are stuck on the surface of quartz grains, then this will have a delaying effect on the dissolution of quartz, Figure 15. The surface of quartz does not come into contact with soda ash. If one is aware of its presence, most of the clay may be eliminated by mineral processing methods. In the worst case, clay can be the cause of quartz–mullite defects in glass. In research comparing sands with and without clay, it was found that sand encapsulated with clay mineral has a retarding action on dissolution.

The presence of clay on sand grains increases the Fe and Al content. Looking more broadly it is tentatively assumed that, not only clay which may pose this problem, sericite, chlorite, mica and glauconite may also give similar results, which are all Fe-bearing minerals.

Limestone, dolomite and feldspars have their own characteristic features. The thermochemical behaviour of carbonates (limestone and dolomite) in the glass melting furnace is believed to be one of the reasons for refractory corrosion. These carbonates yield a wild reaction during the release of $CO_2$, giving off very fine particles to the atmosphere. The amount that flies off is **decrepitation**. The particles do not only go to the superstructure refractories but also to the regenerators as carry-over. The decrepitation ability of carbonates is directly related to the mineral texture. Fine grained crystals have low decrepitation [20].

## 5.5 Refractory Related Defects

All glass productions are made in furnaces lined with various refractories. The melter, the working end and the conditioning section, the forehearths are all lined with different refractories for glass contact and in the superstructure. Refractory choice for different sections of the melter can vary taking into account the campaign life of the furnace. The same approach is applied to other parts, all the way down to the forehearths. During the selection of refractories maximum attention is paid to the fact that the refractory selected should not be a potential source for glass defects. Since there are many refractory suppliers, it is probably ideal to carry out refractory behaviour tests prior to the final decision. If there are in-house facilities this should not be omitted, otherwise this service can be commissioned from other laboratories. The money spent is well worth it.

Glass producers all feel happy if they do not have any problems with their refractories; both for the campaign and any refractory defects that may disturb the production. In reality this is not very likely, there may be defects to some degree, hopefully at tolerable levels. All defects coming from the melter, including the refractory related ones form a background of defects typical of the furnace, which form the daily glass defect intensity.

Production efficiency is about 80–90%, the remaining 10–20% is made of forming defects and defects originating from the melter (batch and/or refractory related).

Sufficient information about batch related defects has been given in the preceding sections. There is a rich literature pool about glass defects in general. Some proposals made for raw materials and refractories have been valuable inputs to these sections for improvement of their products. Due to the improvements made by refractory producers campaigns of 10–15 years are quite easily accessible.

Refractory related defects originate both from glass contact and superstructure. This means that there is a continuous and steady corrosion during the campaign. Although the refractories used in the glass industry are produced with special care, the total volume of refractories used is about 3–5% of all refractory production. The main users are the steel and cement industries.

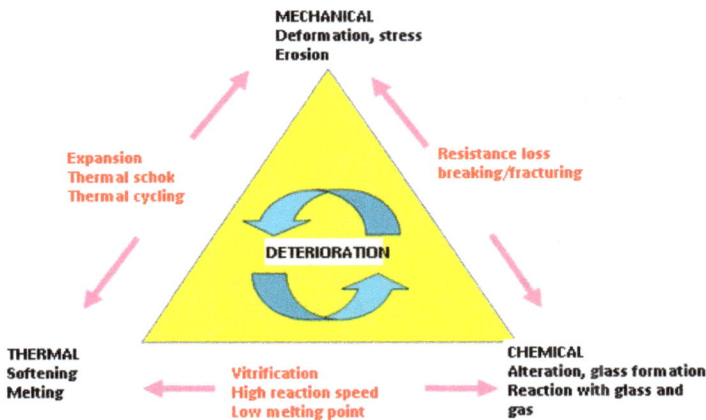

Figure 24: Refractories are continuously subjected to thermal, mechanical and chemical effects. The interactions and alterations which take place are indicated at sides of the triangles.

As seen in Figure 24 the corners of the triangle signify mechanical, thermal and chemical effects, all of which, are active and very much in harmony in refractories used in glass furnaces. The interrelated effects are shown at the sides, the overall effect is indicated in the triangle with the result of refractory deterioration [42]. High temperature and other parameters like glass composition and vapour atmosphere can easily speed up refractory corrosion and erosion, to varying degrees.

The final performance of refractories can be observed at the end of a campaign. It is strongly advised to carry out a post-mortem study at this stage and look for any valuable complementary information which was impossible to reach during the campaign. There is a lot to learn. Make proposals to the project designer.

One can see that there are tons of refractory material worn out in glass contact, most of which has dissolved and disappeared in the glass, with some showing itself as glass defects. Some of these defects were stones, some of which have undergone enough dissolution and may have appeared as knots or cords during glass production.

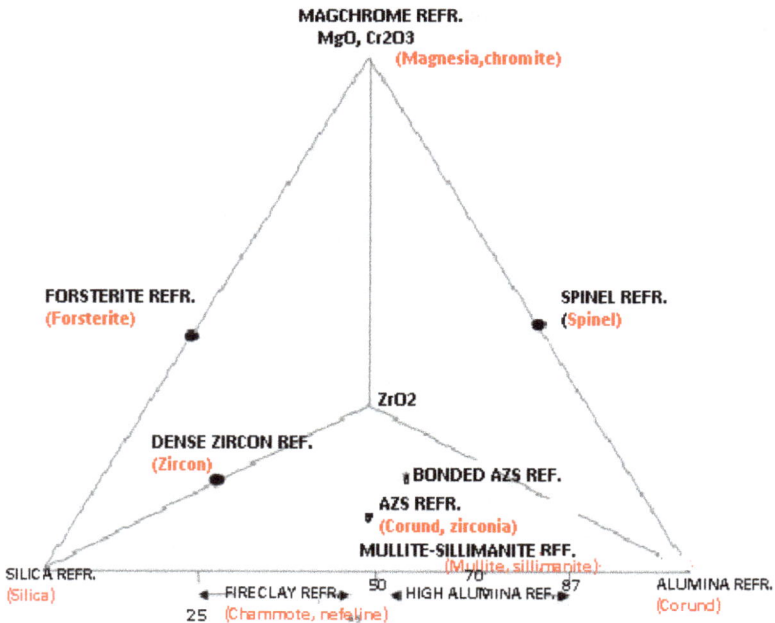

Figure 25: Refractories used in various parts of the furnace are indicated in one diagram. Most of the times defects which originate from a refractory are named after the name of the refractory itself. The names in brackets signify the defects.

Refractories of all kind used in glass furnaces are shown in one diagram, Figure 25 for ease of reference. The refractory types are shown and the related glass defects are indicated in brackets. All these defects are caused, in one way or the other, by mechanical, thermal or chemical reactions.

To identify potential refractories one should build a good collection of refractory samples; which will certainly help to identify the refractory defect. It is also an absolute must that a project drawing showing all the refractory types be available. The defect investigations are made by using several analytical methods (XRD, EMP-EDS, XRF, DTA/TGA, optical microscope, image analyser, etc).

## 5.5.1 Defects Related to Glass Contact Refractories

Defects originating from glass contact are not subjected to furnace atmosphere effects and have not been in contact with flames. These defects develop in the molten glass. There is a continuous impact of erosion due to glass currents.

Accelerated erosion occurs in the sections where the temperature is high, for example near the hot spot or electrode and bubbler blocks. The imprints of upward and downward flowing glass currents have a strong effect on the erosion of refractories, Figure 26.

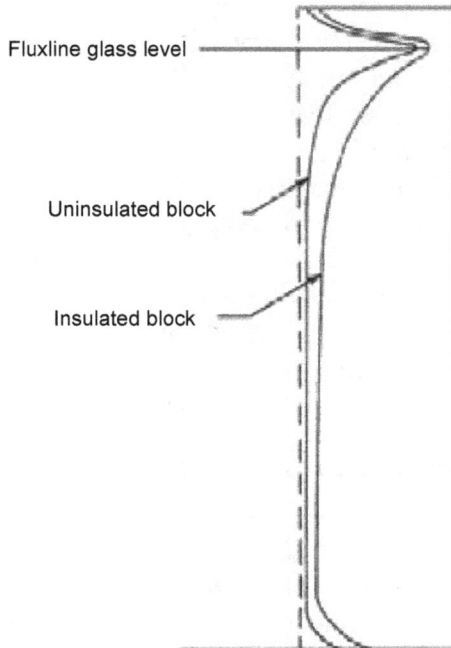

Fluxline glass level

Uninsulated block

Insulated block

Figure 26: The interaction and the result at glass contact refractories. Glass erodes the refractory in several ways. Maximum effect is seen at glass level. It is clearly shown that the effect varies with the presence or absence of insulation. Insulation enhances the extent of erosion.

Bubble release is also possible under these circumstances. Corrosion and erosion of refractories is maximum at the joints. Therefore, minimising the number of joints is preferred at the design stage by selecting optimised dimensions for refractories. Stone defects released from the glass level or below bear identical properties with refractories. If the defects are believed to be related to the glass level, then it would be logical to drop the glass level by small increments. It is more or less traditional practice to reduce the glass level by 5–10 mm/year. These defects are referred to as 'primary'. All facilities (thermo vision, endoscope, optical) available should be used to observe these critical places. If on the other hand, some locations or joints are red hot then suitable cooling should be applied as soon as possible. If corrosion at glass level becomes critical and cooling is not sufficient, then back up blocks, AZS or chromium bearing refractory composition, should be installed; this will prolong the furnace life. If needed a second back up course can also be useful. In general glass contact defects are easier to determine as opposed to those originated from superstructure refractory of the same composition. Defects from glass contact are generally rounded due to travelling in the glass melt, where some dissolution is unavoidable; whereas those originated from superstructure have reacted with the furnace atmosphere whereby a serious alteration takes place. The glass contact defects typically have a solution sac around them. In the low temperature sections of the glass melt, towards the bottom, dissolution becomes extremely low. Even though this is the reality, it is still preferable to cycle the defect in the glass melt as long as possible (extend the residence time). If the defect seem to be concentrated on one side, then one can be suspicious of asymmetric glass currents or they are directed by stirrers.

### 5.5.2 Glass Defects Originating From Superstructure

Considering the conditions under which a glass furnace operates, high temperature and severe atmosphere, it is a typical reactor. Different events are in progression above and below the glass level. Unfortunately, there is no any means of observing what goes on under the glass melt. This is a weak point. However, it is possible to get an insight about the glass currents below the glass level by using mathematical or physical modelling. This information does not allow one to make satisfactory interpretation about glass defect development, however estimations can be made about the manner of travel and trajectories.

For defects originating from superstructure observation is easier. Even though conditions for the observation of some points are difficult, every effort should be made to make the observation of superstructure by all means available. The following points should be iterated, the crown, the tuck stones and the flame geometry (affecting crown, sweeping the blanket, refractory impingement), whether there are potential points for air-ingress (appear dark in colour), furnace pressure; the overall impact all together or individually, Figures 18–20.

### 5.5.3 Silica Defects Related to Superstructure

In soda–lime–silica glass production furnaces (float, container, tableware) the crowns are made of silica refractory. The melter, working end and conditioning ends of the traditional furnaces and the superstructure of flame free zone of float furnaces are all lined with silica refractory. It is such an important refractory type that all features (strong and weak points) of the material must be fully understood. Temperature and furnace atmosphere have serious effects on silica refractory, if there is no direct flame impingement life is relatively tolerable for furnace operators. The enrichment of $Na_2O$ in the vapour atmosphere directly affects the refractory.

NaOH vapours do not affect silica refractory below 1380°C in furnaces where air is used as the source of oxygen. If oxygen is used in the furnace, then the temperature below which the refractory is affected is 1460–1480°C. The maximum vapour impact is seen in the first section, where the batch is charged. Corrosion in this section may show itself in the form of dripping. If there is air leakage from the joints serious corrosion can take place, like rat holes. If the surface of the refractory appears wet then there is flow down from the crown on to the superstructure. These flows may drill the AZS refractory on which it flows. Very frequently it is possible to see silica flow in association with secondary zirconia ($ZrO_2$) crystals. Sometimes the glassy phase may contain $Al_2O_3$ and $ZrO_2$. To prevent this type of silica flowing, it is a widespread practice to use a drip course, along the furnace between the silica crown and the superstructure. The flows are in the form of long drops and are made up of trydimite and crystobalite crystals embedded in the glassy phase. Crystals may be oriented parallel to the flow.

If there is a carry-over of the batch to the AZS refractory then each particle forms a potential point for corrosion. The defect originated in this way is a combination of AZS and silica. On the other hand, silica defects may be related to joint corrosion but also from silica mortar used for repairs.

**Silica frost** is a special lamellar type trydimite defect which shows up in furnaces after 1·5–2 years of the campaign. It disappears from time to time, then crops up again. Silica frost has not been observed in container and tableware furnaces, which may be due to the fact that there is no flame free area in these furnaces. Silica frost is typically trydimite with plate like or lamellar crystals [38,39]. In furnaces where natural gas is used as fuel, it forms on the surface of silica crown and on silica refractories in the flame free zone. At the end of the campaign it can clearly be seen, sometimes even at the conditioning end, the crystals formed here are smaller than those formed at the melter end. If the refractory is rubbed by hand the crystals are easily released because they are loosely attached. Silica frost is a typical recrystallization product, the formation of which is encouraged by the alkali vapours and $SO_3$ in the atmosphere. The first stage is the wetting of the refractory surface, followed by formation of crystals during drying.

Experience shows that if the atmosphere of the refractory zone is diluted with excess oxygen by using special burners, with soft burning style, then the formation can be retarded. The intermittent appearance and disappearance is most likely due to pressure variations in the furnace. Since the density of these crystals is lighter than glass, they float on the surface and appear on the top surface of the ribbon. As this is the very end of the melter, the crystals have no chance to dissolve.

It has not been possible to correlate frost formation with the quality of silica refractory. Different brand and quality refractories are all likely to produce silica frost to some extent.

## 5.6 The Behaviour of $Al_2O_3$–$ZrO_2$ Containing Refractories

Corrosion is unavoidable if flames impinge on the refractories coupled with the effects of furnace atmosphere. The sequence of events starts with the charging of the batch to the furnace and goes on uninterrupted. The maximum effect on refractories can be observed up to port 1. Due to this fact the selection of refractories is highly important, such that they will resist the effect of aggressive and intense alkali environment. Scaling of refractories (1·5–2 cm thick) in this part is quite frequent, and are quite likely to fall on the batch blanket. Formation of scaling over and over is quite possible. This type of problem is frequently seen in float furnaces, especially in the suspended doghouse wall (typically known as A-wall effect). If bonded AZS is used in the back wall the coarse alumina crystals do not have a chance to dissolve and they appear as stone defects. All bonded AZS refractories are likely to undergo scaling and give rise to stone defects in this part of the furnace. The zirconia component of bonded AZS refractories is very small in size (20–30 μm) therefore they do not appear as defects because these crystals dissolve in the glass melt.

The intensity of superstructure dripping at the tuck stones increases towards the hot point. This drippings can also be seen at the port arches. If all the dripping finally flows into the glass, it is fair to assume that they will give rise to many vitreous defects [43].

The activity of these droplets is related to temperature. If the temperature is high enough they will easily flow into the glass, if not they will stay fixed as they are. The homogenisation of these droplets in the glass melt is highly difficult, since the liquidus temperature of the droplets is about 1400°C, and the liquidus temperature of the melt is in the 930–1000°C range. According to physical model findings, these droplets, once they are in the glass melt, travel along the sidewalls within 150 cm from the right and left of the furnace.

They are directed by the glass currents. Some of these after arriving at the target wall then return towards the hot point (like a swimmer in the swimming pool) where they find a new opportunity to dissolve. Some of the droplets go through the waist with the forward glass currents. In this case there is no dissolution and finally they appear as knots and cords in the product. At the highest temperature area the glassy material takes the form of curtain flow, in a continuous manner.

Some types of electro cast refractories especially which have 31–32% $ZrO_2$ in their composition and release their glassy phase easier. In this case it is highly logical to use more resistant refractories in critical places. At cold repair it is possible to see that about 1–2 cm depth of the refractory has gained a sponge like texture, almost with no glassy phase at all, leaving only zirconia crystals behind. This depth may reach 2–3 cm near the hot point and around port entrances. The main components of AZS refractories are zirconia and corundum crystals. In these critical places corundum crystals dissolve in the glassy phase of the refractory and leave behind a sponge like structure [44].

In the droplets thus formed there is a high $Al_2O_3$ (<20%) and less $ZrO_2$ (<5%), Figure 27. Knots and cords of this chemical composition in soda–lime–silica glasses, with no doubt originate from these glassy droplets.

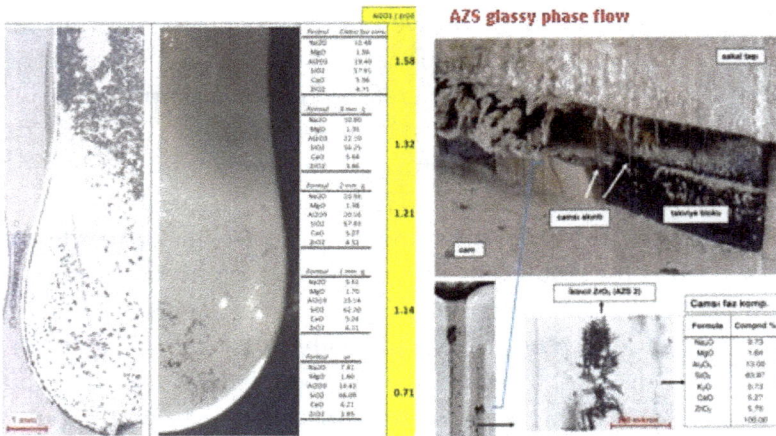

Figure 27: A section of a drop with high secondary zirconia crystal content near the tip, at the root of the drips the size gets larger. In a drop of about 10 mm the $Al_2O_3/ZrO_2$ nearly doubles. There is less alumina near the tip. In a cold repair it is possible to collect these glassy phases.

## 5.6.1 Some Facts about the Glassy Phase of AZS Refractories

AZS refractories of high corrosion resistance are traditionally used in critical areas within the furnace. These refractories characteristically have 15–20% glassy phase, which oozes out of the refractory and drips into the glass melt to give defects like knots and cords.

In some of the literature it is pointed out that the $Al_2O_3/ZrO_2$ ratio of the glassy phase may indicate where the drops originate. Numerous analyses made in our laboratories clearly indicate that this is not correct, because the referred ratio of the drop varies remarkably in the drop itself.

To explain the importance of the glassy phase, let us assume that the refractory in this case is standard AZS, which contains about 6 kg of glassy phase in a 1 m×1 m×1 cm slab, by simple calculation. If we further assume that a typical knot is of 1 mm diameter sphere, then there will be about 400 knots in one free falling drop, Figure 28.

Some Facts about AZS Refractories Glassy Phase

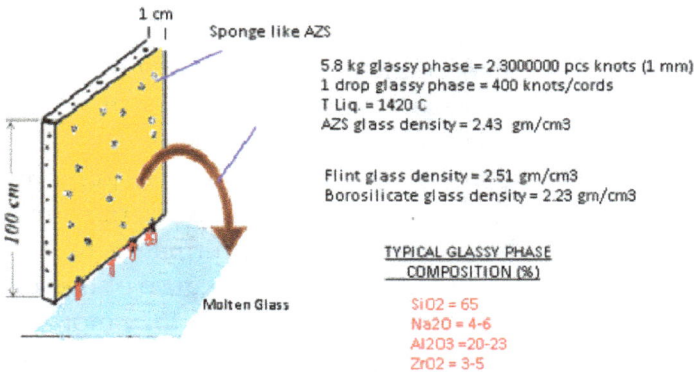

5.8 kg glassy phase = 2.3000000 pcs knots (1 mm)
1 drop glassy phase = 400 knots/cords
T Liq. = 1420 C
AZS glass density = 2.43 gm/cm3

Flint glass density = 2.51 gm/cm3
Borosilicate glass density = 2.23 gm/cm3

TYPICAL GLASSY PHASE
COMPOSITION (%)

$SiO_2 = 65$
$Na_2O = 4-6$
$Al_2O_3 = 20-23$
$ZrO_2 = 3-5$

Figure 28: The diagram is intended to show what happens at the superstructure and the analytical values obtained. An attempt has been made to calculate the number of knots that may arise from one single drop (400 knots). Recalling the impact of knots on glass quality, the above assumption poses a dramatic picture. The effect of temperature of a defect of 1mm diameter on dissolution is as follows:

|  | 1200°C | 1300°C | 1400°C |
|---|---|---|---|
| **Dissolution time of 1 mm thick defect** | 600 h | 140 h | 30 h |

By using mathematical modelling the temperature distribution of the glass melt turns out to be of rather low order, it is only a small depth of glass melt which can be rather active in dissolving defects, the larger depth of glass melt is well below 1400°C. For furnaces (float, container, tableware) with 1580°C crown temperature the volume of glass which is above 1500°C is approximately 2% in float, 1% in container and 5% in tableware, Table 14. These figures clearly show that the thermal efficiency of glass furnaces is highly low.

| | Volume of glass at different temperatures | | |
|---|---|---|---|
| Furnace type | >1400°C | >1500°C | >1550°C |
| Float | 12% | 2% | <1% |
| Container | 13–20% | 1% | <1% |
| Tableware | 74% | 5% | <1% |

Table 14: Approximate values of the volume of glass at different temperatures is shown for float, container and tableware furnaces.

The above iteration necessitates for best results of dissolution that the glassy phase content of the refractory should be as low as possible, and should resist to coming out of the refractory. These points should be on the agenda of priorities by the refractory producers. No doubt furnace operation of glass producers is also equally important.

In general, at the design stage of the furnace, all the combinations of refractories must be iterated for best performance. It is never wise to use an average quality refractory just because of the lower cost. This will imperil the quality, and this turns out to be more expensive. Whether it is the initial cost of investment or saving energy, it is the quality of glass that matters first. If there is no good quality glass then there will be no business.

The material which give rise to knots and cords is essentially the same. The extended tail of a knot is in the form of cord. Depending on the viscosity and the ability to dissolve both defect may develop. Knots and cords can be present to varying extents in all types of glasses (float, container, tableware). Those which originate from the melter can invariably be related to electro cast AZS refractories, which characteristically contain $Al_2O_3$ and $ZrO_2$. However, it is possible cords or knots may be enriched in $Al_2O_3$ and contain no $ZrO_2$. These are called as **aluminous knots and cords**, which will be dealt with in detail later.

$Al_2O_3$ and $ZrO_2$ bearing vitreous defects are frequently encountered in glass production. According to classical opinion these defects originate from the superstructure and flow into the glass melt, subsequently reaching the bottom levels of the glass tank due to their higher density. According to our research, over at least two decades, these defects were found to be lighter than glass and therefore were transported near the surface of the glass dispersed in the forward flow glass current. Put more bluntly the density of $Al_2O_3$ and $ZrO_2$

bearing material is not heavier than glass. They are not allowed to sink to the bottom due to the forward flowing current. The continuous linear defects seen in float glass, the cat scratches in container glass and cords in tableware are all of the same origin and behave in a similar manner.

In hand produced glasses cords are not rare, they are physically similar, and disturb the aesthetics of the ware. These type of products are easily refused by the customer, there is no tolerance at all. The origin may be the melter and conditioning end refractories they may be siliceous or aluminous in composition. These furnaces are dimensionally quite small, therefore flame impingement on refractories is unavoidable. It is advisable to produce burners suitable to the span or length of the furnace.

### 5.6.2 Vitreous Defects in Float Glass

Cords or continuous defects in float glass lie parallel to the pull. The thickness of these lines are about 20–50 µm and they contain micro-knots on the cord line. Very frequently it is possible to see secondary zirconia crystals in the micro-knots. These finding alone proves that these must be related to AZS refractory. The pattern and frequency of these micro-knots gives the impression of a 'stitching line', Figure 29. Continuous cords appear intermittently in float glass production, therefore, the total time of appearance is measured.

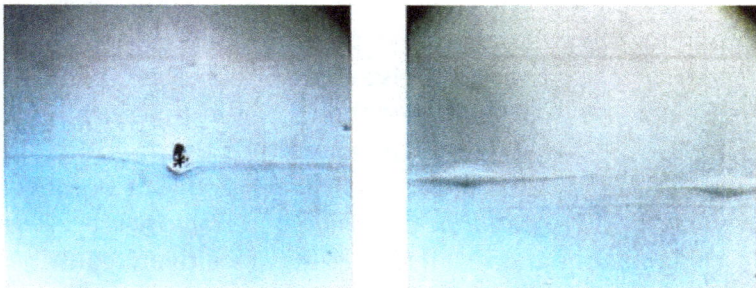

Figure 29: Continuous cord in float glass showing secondary zirconia crystals, related to AZS refractory. The 'stitching pattern' is shown on the right.

Some features may be interesting: about 70% are found in the top half of the ribbon thickness, and they seem to be gathered on the left and right sides, and noticeably less in the centre of the ribbon, Figure 30, therefore they are not likely to return to the hot point. These thin lines are more visible in thin glass as compared to thick ribbon.

GLASS CURRENTS

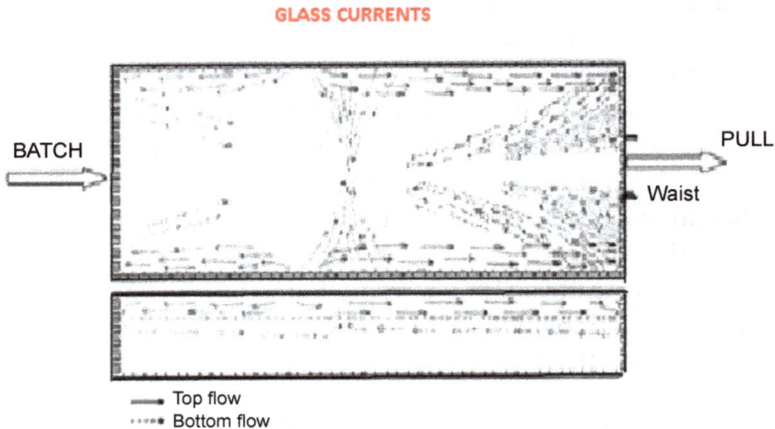

Figure 30: In a physical model study of a side-fired furnace the superstructure drippings appear to travel near the sides within about 140–150 cm. Some of these return back to the hot point, the majority tend to position themselves in the forward flow near the surface. In glass sheets a similar distribution has been observed, these cords are more concentrated near the sides and less in the centre [7].

The presence of these lines creates some serious quality problems especially in automotive and mirror glass. To prevent these is not easy, as they are the corrosion products of AZS refractory due to high temperature and fierce atmosphere conditions.

If the refractory has any negative contribution, then the only solution is to increase their residence time in the furnace by all means available. If the temperature is increased then the viscosity of glass and glassy phase will be lowered, which is good for homogenisation, but on the other hand the corrosion of refractory is accelerated.

Electric boosting and/or a bubbler may positively contribute to the problem as these do not increase the superstructure temperature. The stirrers in the waist forehearth canal have a dominant effect on reducing the size of cords by stretching and folding. As these defects travel in the forward flow, probably some developments can be made to the stirring style [45, 46].

**A deep waist cooler** keeps the glass melt longer in the furnace, thus improving homogeneity. A similar effect can be obtained if the temperature difference between the hot point and the doghouse is increased. As mentioned before, it would be ideal for the glassy phase not to come out from the refractory. Some success can be obtained if the refractory texture is compact. It remains for the furnace operators to have a tight control on the flames.

The flame velocity should not exceed 8 m/s, and the flames should not sweep over the batch blanket which causes carry-over to the refractory resulting in accelerated corrosion. Of course excessive insulation of the superstructure is not favourable in this case. It is worth mentioning here that the decrepitation of limestone and dolomite create the same effect.

Continuous cord defects in float furnaces has been a point of concern through the years, and numerous microanalyses have been made to put a diagnosis to the problem. The cord analyses in container and tableware products seem to gather around an area, but in float glass the distribution of cord analyses seem to cluster in two well defined areas. The $Al_2O_3$ and $ZrO_2$ distribution is highly interesting and needs explanation. The approximate range is given below, Figure 31.

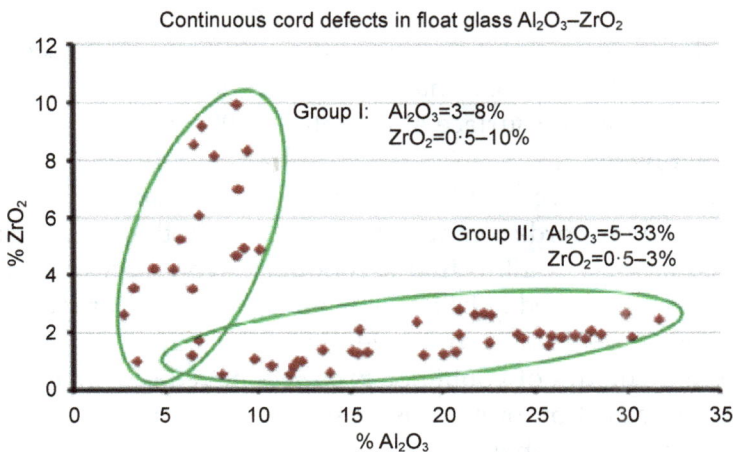

Figure 31: The results cluster in two areas, which needs some convincing explanation.

| | % $Al_2O_3$ | % $ZrO_2$ |
|---|---|---|
| Group I | 3–8 | 0·5–10 |
| Group II | 5–33 | 0·5–3 |

It is quite possible that there may be two sources or in different furnaces diffusion mechanism on vitreous defects may differ. In the refractories where zirconia is about 32% and 41% the glassy phase composition is closely similar, which is not enough to explain this trend, Table 15.

|           | 32% AZS refractory | 41% AZS refractory |
|-----------|--------------------|--------------------|
| Na$_2$O   | 5                  | 5                  |
| Al$_2$O$_3$ | 20–21            | 18–19              |
| SiO$_2$   | 67–68              | 69–70              |
| ZrO$_2$   | 4–6                | 5–6                |

Table 15: The glassy phase compositions of AZS refractories with 32% and 41% zirconia content. It is not possible to differentiate cords which originate from these type of refractories.

### 5.6.3 Vitreous Defects in Container Product

Vitreous defects in container glass products are widely known as 'cat scratches'. They extend in a parallel manner from the top to bottom, frequently many lines in one zone. The name recalls the appearance of cats claw imprints. If it is examined carefully it will be seen that they extend between shear marks. The thickness of cords are 10–50 μm and are near the surface at a depth of 100–150 μm, which can be felt by finger.

Sometimes they cover a large area, in a rather narrow zone. The majority of these cords contain Al$_2$O$_3$ and ZrO$_2$ in their composition and it is not rare at all to find secondary zirconia crystals on the cords. This is assumed to be a firm evidence that they are related to the glassy phase of AZS refractories, Figure 29, [5].

Different colours of container products have been investigated thoroughly based on numerous analyses. The range of Al$_2$O$_3$ and ZrO$_2$ are shown in Table 16.

|        | % Al$_2$O$_3$ | % ZrO$_2$ |
|--------|---------------|-----------|
| Flint  | 3·6–12·4      | 1·0–3·2   |
| Green  | 5·9–10·7      | 1·0–3·4   |
| Amber  | 3·5–6·2       | 1·9–2·6   |

Table 16: The Al$_2$O$_3$ and ZrO$_2$ contents of cat scratches encountered in different colour container products.

The presence of cat scratches has been overlooked for many years, but of lately the customer has become more critical, and quite often has been a cause of rejection.

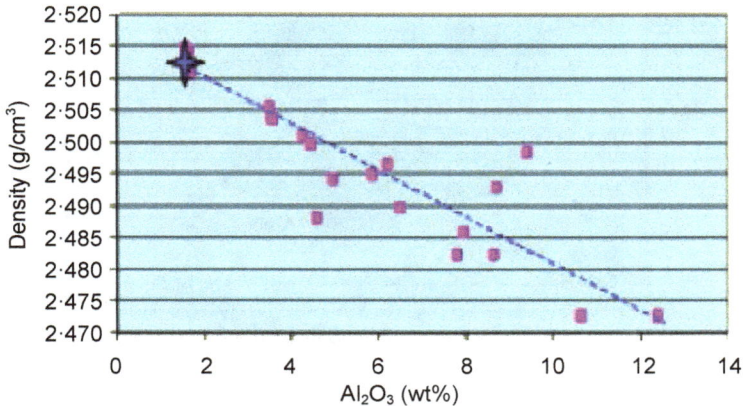

Figure 32: The density of catscratches is clearly lighter than the base glass (blue star). Density decreases as $Al_2O_3$ increases.

As mentioned before the density of these defects is lighter than the base glass [8]. Therefore they are not expected to sink to the bottom, Figure 32. There is a wealth of literature on cords and cat scratches, in the past 50–60 years, all of which have exclusively commented on the fact that the material which leads to vitreous defect is considered to be heavier than glass. Nevertheless, there has not been any experimental or theoretical study made on this topic to prove the point put forward. It is presumably due to this confusion that no solution has been reached up to now. A direct implication of this belief is that drip-draining to get rid of this material has never been successful. That is understandable because the vitreous material is not heavier than the glass. Added to this there is another point which is rather interesting; that is the stirrer itself may give rise to some complications. The vortexes formed in front of the stirrers can release some material intermittently, effecting the quality periodically. It is the author's opinion that the stirrers should be improved to avoid this problem.

The variation of analytical data of cat scratches is observed to be minimum in amber glass products, as compared to flint and green colour products, Figure 33.

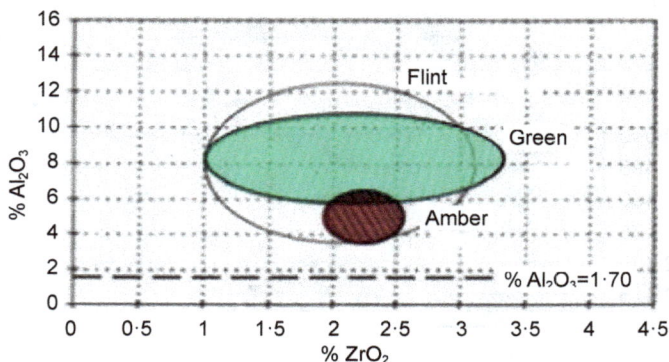

Figure 33: The spread of $Al_2O_3$ and $ZrO_2$ in cat scratches seen in container products. Note that in amber glass the spread is limited. The $Al_2O_3$ level of base glass (1·7%) is also shown.

The most practical approach would be to minimize the formation of glassy phase. Once it drops into the glass melt it is very difficult to homogenise. This is due to the fact that about 80% of the volume of glass is below 1400°C, not enough for efficient homogenisation. What is more the liquidus temperature of cord making glass is about 1400°C, while that of the base glass is about 1000°C.

The control of flames is important in that they should not cause any carry-over to the refractories. In end-fired furnaces maximum care should be taken to avoid flame impingement to the front wall, which is usually of AZS refractory material. Any drops from here into the glass near the throat will never have any chance to dissolve. Owing to this fact the insulation of the front wall should never be excessive.

If there is a bubbler in the furnace it may be useful for defects which come from the first half of the furnace, but there will be no positive contribution for those originating from the front wall. Stirrers in the forehearths may help in stretching and folding of cords, giving smaller, tolerable sizes [48]. As mentioned before stirrers may create some problems, for which mathematical modelling is necessary. For example, the defect loaded glass may be pushed to one side of the forehearth.

Knots are also abundant in container products, the same iteration as above is valid here too. To increase the residence time of the defects will be advantageous. The flames as they leave the burner block refractory may touch the periphery, in this case some large size knots may form.

### 5.6.4 Knots and Cords in Tableware Products

In all glass production type's vitreous defects are of serious concern for the producer. For prestigious products of tableware these are not tolerable at all, therefore there are strict quality control limits. For very thin products not only cord defects, but none are allowed. Low iron products are even more critical. In automatic production cord limits are not very strict if there are optical decorations. In fact this is one way to suppress or hide the impact of defect. The composition of cords are highly variable ($Al_2O_3$=1·3–11%, $ZrO_2$=0·5–9%). Sometimes $ZrO_2$ may be absent, in this case the defects are high $Al_2O_3$ containing knots and cords. Vaporisation of tableware glass is more effective, superstructure aluminous refractories are affected extensively. The glassy phase formed on the surface of refractory may flow into the glass in the form of droplets. The channel refractories and the cover blocks are made of sillimanite or mullite, both of which are seriously affected by vapours from the glass surface. It is worth to remind that if feldspar raw material is coarse it may give similar defects.

If high alumina defects come from one line or more, then one should be suspicious of the zone separation block touching the glass. All observation means (endoscope, binocular, etc) should be used to verify this. In the project stage these separation blocks are designed to be as near to the glass as possible (5 mm). During the heating up this distance may change to the extent of touching glass. If the zone separation is in contact with the glass then defects are continuous and unavoidable. The only remedy is to lift up the block to a safe height. It will be a good operational practice to check the zone separation blocks after glass production starts.

Vitreous defects in tableware furnaces are related to melter refractories. There are a lot of droplets at the tuck stones, which are all candidates for knots and cords. On the other hand the target wall which is usually of AZS material, is very critical source for defects. It is worth repeating again that the insulation level of target wall should be kept at minimum. Because of high alkali content ($Na_2O+K_2O$) and the very low $Fe_2O_3$ level the dissolution ability of this glass, as compared to float and container glass, is very much higher. More than 70% of the glass volume is above 1400°C.

To keep the problem within tolerable quality limits it is widespread practice to use stirrers in the forehearth channels [47]. They are universally accepted to be useful, but sometimes a pulsing attack is also noted.

The classical stirrers are based on the principle of pumping from bottom to top, which assumes that the cords are heavier and travel from the bottom. In our research it was established that the cords are lighter than the glass, and flow near the top surface, Figure 33. Therefore stirring from top to bottom would be more logical. By physical modelling it was observed that homogenisation is much better. Dip-draining in the forehearth channels has no positive contribution, and never has been a solution for the problem. It can easily be said that the volume of glass thus drained is a loss, which undoubtedly amounts to some weight, equivalent to many products.

## 5.7 Bubbles in Glass and Their effect on Quality

Following the charging of batch to the furnace, thermochemical reactions take place immediately. It is appropriate to remember that to produce 100 kg of glass the batch needs to be 120 kg. Therefore, after all the conversions and reactions are achieved 20 kg of gases are released from the batch, the majority of which are carbonates, sulphates, moisture, and nitrates. These gases are first in the furnace atmosphere and subsequently released as emissions to the environment. The glass formed after reactions has a high viscosity, which makes life difficult for the bubbles to escape from the glass melt. The release of gases from the glass (fining) takes a long time; not only this but at the same time work is neede to homogenise the glass melt, which is so important for good quality glass [12]. The **batch free time** is how long it takes for the last component of batch to dissolve. The energy required to achieve this is about 90% of that used in glass melting. Although homogenisation and fining together require a specific time (residence time), the energy required is only 5–10% of the total. Like batch free time, there is also **seed free time** for bubbles or seeds to reach a level which is acceptable for the quality of glass. Residence time in float glass furnaces may vary from abnormal values of 4–8 h to normal values of 70–120 h. If the forward flow current is variable in behaviour then there may be some short cuts for some glass, giving 4–8 h residence time. In this case the quality of glass is well off the limits, and the fining stage is a failure.

The seed levels that can be tolerated in float glass are extremely low compared to container and tableware glasses. In all the glass types their presence is in close relation with temperature, seed free time, chemical composition of glass, the reducing and oxidising conditions of the furnace and the level of refining agents. During the melting process some of the gases are released from the glass melt, while some are dissolved in glass and yet others are found as gaseous inclusions. There are two fundamental mechanisms for gas release from glass:

- **Fining:** The size of bubbles increases as it rises in the glass melt. This behaviour is favoured at high temperatures.
- **Refining:** As bubbles dissolve in glass their size gets smaller and they finally disappear. As opposed to fining this behaviour develops as the temperature of glass decreases.

The behaviour of bubbles in a viscous medium is guided by Stoke's Law:

$$V_s = \frac{2g\Delta\rho r^2}{9\eta}$$

Where, $V_s$=rising velocity of bubble, $g$=gravity, $\Delta\rho$=density difference between glass and bubble, $r$=diameter of bubble, and $\eta$=viscosity of the glass.

Looking at the equation it is easy to infer two points; the first is that $\Delta\rho$ is not affected easily, the second point is that the most effective parameter is $\eta$. It is possible to adjust viscosity by changing the composition of glass (increase alkali and temperature).

In soda–lime–silica glasses the dissolved gases are expelled by using $Na_2SO_4$. The solubility of $SO_3$ depends on several parameters;

oxidising and reducing character of furnace atmosphere, residence time and temperature and glass composition.

As sulphate reacts with $SiO_2$ at about 900°C, gases like $SO_2$ and $SO_3$ are evolved. These gases play a primary role for the refining of glass.

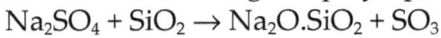

$Na_2SO_4 + SiO_2 \rightarrow Na_2O.SiO_2 + SO_3$

$SO_3 \rightarrow SO_2 + 1/2O_2$

The overall reaction is;

$Na_2SO_4 + SiO_2 \rightarrow Na_2O.SiO_2 + SO_2 + 1/2O_2$

This reaction will be completed before 1450°C. $SO_3$ containing bubbles coalesce with one another so that their diameter increases, which is an advantage for easier ascension. In the meantime as temperature falls the small bubbles are dissolved in glass melt.

### 5.7.1 Trouble Shooting for Seeds and Bubbles

The gaseous inclusions in glass affect the quality of products. Their number and size, the distribution on the ware, whether they are near the surface or not, all affect the fate of the products. To find the cause of bubbles it is necessary to know the parameters (pull, temperature, cullet ratio, etc) before the breakout. The microscopic properties (deposits, size, etc) and if there is gas, analyses should be known.

**Blister** is the common name used for bubble and seed according to International Commission on Glass definition; in which the following size classification is proposed: bubble>0·2 mm, seed<0·2 mm in all glass types.

### 5.7.2 The Evolution of Gaseous Inclusions

At the end of the melting process it is assumed that the following are potential sources for bubble

* The thermochemical reactions of raw materials,
* Air entrapped in the batch,
* Interaction between glass and refractory,
* The oxidising and reducing character of furnace,

The analytical characteristics of bubbles are given in Table 17.

Bubbles can be incorporated into glass melt at any stage and this may well continue until the forming.

| | |
|---|---|
| • **$CO_2$ Major** | • **$SO_2$ Major** |
| • Refining>$SO_2$ | • Reboil>$CO_2$, $O_2$ |
| • Dissolution of devit>CO, $N_2$ | • Reduction of sulphate>$CO_2$, |
| • Carbonate decomposition>$N_2$, Ar | $N_2$, $O_2$ |
| • Metal contamination>CO, $N_2$, $O_2$ | |
| • Refractory corrosion>$N_2$, $CH_4$ | • **$O_2$ Major** |
| • Batch decomposition, fuel>$N_2$, Ar | • Electrolysis |
| | |
| • **$N_2$ Major** | • **$H_2$ Major** |
| • Air entrapment>$CO_2$, Ar | • Fe contamination>$CO_2$, CO, |
| • Nitrate>$O_2$ | $N_2$+/-$H_2S$ |

Table 17: Analytical characteristics according to
major and secondary gas species in bubbles.

Some of the common causes are as follows:
- Insufficient refining agent
- Insufficient cullet
- Low temperature
- Batch reaching the target wall
- The swaying of batch to right or left during inversion
- Segregation in the batch
- High pull and insufficient residence time
- Organic or metallic contamination of raw materials

If the temperature is high when the batch is charged into the furnace, then a high viscosity liquid formed on the surface of batch, which prevents the release of bubbles. As the chemical reactions take place the expanded gases find a way to escape. At this stage foam starts to form. This is an obvious disadvantage for what is to follow later. In side-fired furnaces to optimise the distance from doghouse to port 1 helps to minimize this effect.

Various water coolers used in the furnace should be checked regularly for water leaks, otherwise seed breakouts are unavoidable.

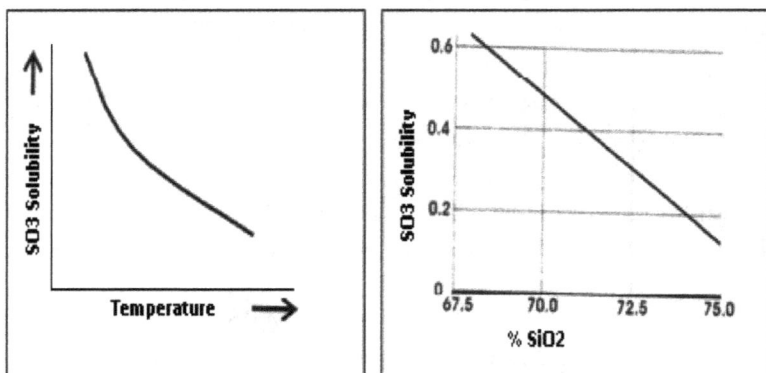

Figure 34: The $SO_3$ solubility in glass increases with decreasing temperature; similarly, solubility decreases with higher $SiO_2$. The doghouse temperature is important for the retention of $SO_3$.

Gas solubility in glass decreases as $SiO_2$ content increases. Therefore, the value of $SiO_2$ in the chemical composition is critical, Figure 34. Gases released at high temperatures can find opportunities to dissolve in the later low temperature parts of the furnace. The efficiency of this depends on the type of gas, and the chemical composition of glass. The increase of alkali favours this reaction. Gases dissolved in glass can be released when they confront a temperature increase. Bubbles released in this way are known as **reboil bubbles**. Temperature increases can be seen in conditioning section and the forehearths. A 10–15°C increase in temperature will suffice to trigger reboiling. $SO_2$ is the gas which dissolves easiest in glass, where Na-sulphate is used as refining agent. These are known as reboil bubbles and $SO_2$ is the major component, Table 18. Glass composition has also an effect on the reboil tendency, for example amber glass composition has a high risk for reboil.

Bubbles related to **electrolysis** are not rare. As the glass is pulled towards the forehearths, there are different sections of the furnace which have different metals (gob shears, level measuring devices, thermocouples, electrodes, water coolers, stirrers, etc). Some of these metals may not have enough earth grounding, which leads to DC development, an essential step for electrolysis. Metals of different composition touching each other should be isolated and grounded.

At about 600°C the glass itself becomes electrically conductive. If a contact with glass is suspected it is advisable to apply a reverse polarity DC potential, which may be a remedy. Electrolysis bubbles contain $O_2$ at the highest level with some minor $CO_2$ and $N_2$ in addition.

## 5.8 Some Problems around Throat, Waist and Conditioning

The melter in end-fired and cross-fired furnaces is connected to the conditioning section, where forming operations take place, by throat and waist, respectively. Almost all glass in the melter passes from the throat and waist. This is a continuous operation until the end of the campaign, therefore as the glass passes through these narrowed sections its velocity and flowing rate affects the refractory structure more than any other part of the furnace. It is due to this that refractories of high corrosion resistance (about 40% $ZrO_2$ containing AZS refractories) are widely used in these critical locations. Depending on the colour characteristics of the glass sometimes even more resistant refractories ($Cr_2O_3$ containing) can also be used.

In end-fired furnaces the **throat** is the main structure which may be the main cause (40%) to stop the furnace. Therefore, special measures like air and water cooling can be applied here extensively. Blocks used in the throat may have a molybdenum insert to increase their resistance. The flow character of glass in the throat is turbulent, therefore erosion of refractories is unavoidable. Similar precautions are valid in the waist area of cross-fired furnaces, although refractory erosion is less as compared to throat.

Foam accumulation around throat and waist is not rare [41,42], Figure 20. The agglomeration of coal in the batch may give rise to serious bubble problems that may extend up to this point. Rundowns originating from the target wall, and those droplets which fall from the section joints can accumulate in front of throat or waist. The density of the accumulated material is lighter than glass, therefore they can remain here for longer periods which causes quality problems. This part of the furnace should be on continuous surveillance.

The **waist coolers and stirrers** are weak spots for water leakage, which cause bubble breakouts, therefore they should be monitored on regular basis. The furnace side of the waist coolers is a potential area for accumulation of bubbles. They may be swept forward from below the coolers with the forward flowing current, reaching the conditioning zone [48–50], Figure 35.

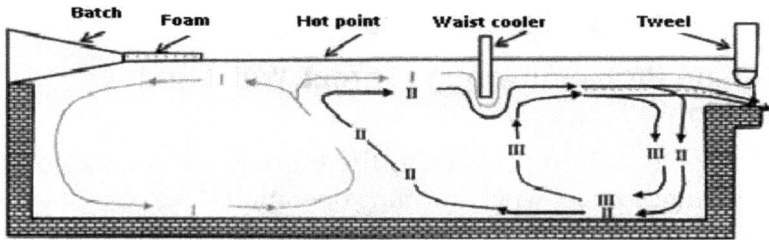

Figure 35: Foam formation is seen first at the end of batch then around the hot point. If any foam bypasses the hot point it will accumulate in front of the waist cooler, which at times may force its way to the conditioning zone passing under the cooler.

The quality aspects of the molten glass (seed level, homogenisation, continuous defects) are closely related to the waist cooler. As the depth of the waist cooler increases the energy of glass will stay on the melter side, as this will increase the residence time quality will be positively affected [51,52]. When the waist cooler is inserted in the molten glass, a cover of devitrified glass of 2–2·5 cm thickness is formed on the cooler, in a short period (12–13 h). When taken out from the glass the surface facing the melter may show some intricate features.

**Stirrers** are another set of important tools in the glass technologist's hands [47]. The shape of the stirring blade may vary. The rotation speed should not disturb the laminar flow, therefore 6–8 rpm is normal. Continuous defects, cords, and knots are reduced in size with the action of stirrers. The stirrers usually are activated by one motor. If there is some discrepancy in their rotation they may touch the cover blocks, care should be taken to avoid this.

If there is water leakage in the stirrers, this is a potential source for bubbles. Condensation of vapours may cause the formation of Na-sulphate on the metal structure of the stirrers, which will give sulphate gall defects. Fibre based materials can be used for insulating purposes of the structure, even this may form defects.

The activity of stirrers individually may be advantageous in some cases [53]. For instance, the extreme stirrers may be used with different rpm. If there is a high concentration of defects at the right and left it may be useful to increase the velocity. As the viscosity of glass may be higher than the centre, then this may also necessitate 1–2 rpm higher. Based on the findings of mathematical modelling the efficiency of stirrers is not the same.

Stirrers like the waist coolers are covered by devitrified glass of 2–2·5 cm thickness. If the water supply deteriorates (flow rate, hardness, temperature, etc) this devit layer may fall off, which creates problems, if there is a leakage then bubbles may occur.

After passing through the waist (float) and the throat (container) the glass melt finds itself in the conditioning zone, where in the real sense of the word glass is conditioned. In float furnaces the conditioning zone is between the melter and the tin bath. The melter entrance is the waist, the tin bath side is known as the canal, which allows the passage of conditioned glass to the tin bath without disturbing any gained properties. The stability of temperature of glass at the canal is very important. One may say this is one of the most critical points in float glass production. To achieve this a good quantity of air is blown into the conditioning zone. No particles should be present in the blown air, otherwise defects on the top surface of glass may occur.

The filters on the air channels should be cleaned or renewed periodically to avoid any particle effects. There should be no moisture in the blown air.

The blown air should not be directed at the surface of glass, the blowers should be angled upwards. If blowing affects the surface of the glass some cooling may result, which in turn may cause some inversion of glass.

# CHAPTER 6

## 6.1 Tin Bath

Tin bath is the most critical part in float glass production, where forming operations are accomplished [54,55].

The canal at the end of conditioning connects the tin bath. The glass flow is adjusted by the fused silica tweel which is in the canal. It is inserted in a special slot, which has up and down action. If the glass flow or pull is to be reduced the tweel is lowered down in the slot; for increasing the pull the reverse action is executed. The tweel resembles the set up in a dam. If there is a problem related with forming operations, then the first thing to do is to lower the steel tweel and stop the flow of glass. The steel tweel is normally lifted and is located towards the conditioning end.

Float glass technology was used commercially in 1967 for the first time, after ten years of patient and expensive research. The aim was to get an optically perfect surface in the glass. This aim has not changed since as the best quality of glass is produced by this technology.

A tin bath has some specific parameters which are universal to all float lines. The tin bath is made of narrow and wide sections, where the width of narrow section is variable, whereas the width of wide section is 280 inch (7·112 m) and is fixed. The length of a bath is divided into bays of fixed dimension that is 10 ft (3·048 m). The width of the narrow section may vary. The number of bays in a bath are related to the pull.

> *According to Pilkington practice:*
> *Number of bays=pull (ton)/31·5*
> *For 700 t/d, no. of bays=700/31·5      >22·2 bays*
> *For 600 t/d, no. of bays=600/31·5      >19 bays*

The width of the ribbon is determined by the width of the narrow section, this value is not fixed and it depends on the requirements of

glass in the market. Several criteria need to be discussed, irrespective of the bath size, prior to the designing of tin bath.
- Bath dimensions
- Glass depth and canal dimensions
- Spout dimensions and installation features
- Wetback and restricter tile geometry
- Tin depths and barriers
- Roof heating configuration and capacity
- The dimensions of exit and LOR
- Atmosphere conditions
- Specification of materials used

The temperature of conditioned glass entering the bath is 1100°C for flint glass, and slightly higher for coloured glass. In any case, the temperatures of glass, flint or coloured must be stable. Only during problem solving instances (+/-) a couple of degrees of temperature change can be allowed. No atmosphere should be allowed to pass the ceramic tweel towards the tin bath.

Tin bath atmosphere is a mixture of $H_2/N_2$, 5–10% $H_2$ and the rest is $N_2$. The quality of the molten tin in the bath is preserved with the atmosphere. Glass with 1100°C temperature is poured into the tin bath with the aid of lip stone (=spout). The pouring of glass into the bath is very much like the flow of honey, extreme care must be taken not to disturb the poring pattern of glass.

This action in this part is the most critical stage of float glass manufacture, if this is achieved with success then the rest is likely to follow accordingly. The pouring action of glass into the tin bath is the most critical point of float technology. Once the glass is on the molten tin, it starts to float. It then travels in the bath of reducing atmosphere until the exit. Before the glass enters the tin bath it is under an oxidizing environment.

Due to the fact glass floats on molten tin the name 'float' has rightly been assigned. The melting furnace and the conditioning end are not necessarily related to float technology as such. The defects originating from the melter and the conditioning end have nothing to do with the float technology. The basic requirement is to have no defects at all, if possible, coming to the tin bath. As it is universally accepted float glass is the best quality of flat glass, provided that there are no incoming defects and also have the least level of defects inherent to the bath. If this is satisfied glass of optic quality is likely to be produced.

## 6.2 Glass Pulling Systems

In order to be able to pull the glass there are several tools available, for instance suitable coolers and heaters are essential. Top roll machines, water fences and carbon pushers are important accessory systems. To obtain the required glass temperature flat coolers are used in the hot section, whereas in the narrow section bank coolers are used, Figure 36.

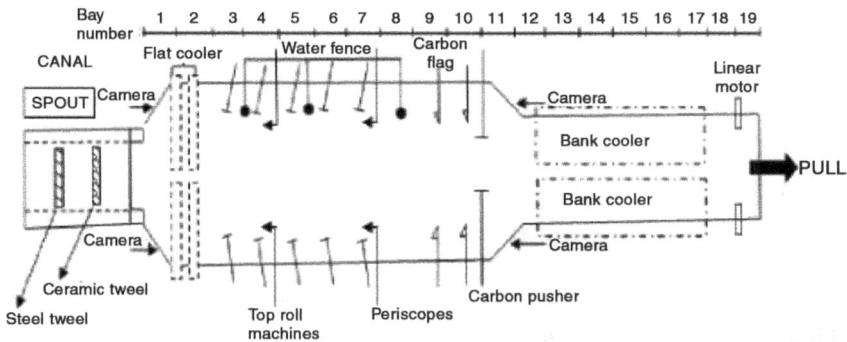

Figure 36: A general view of tin bath, all the subsidiary equipment used for forming are also indicated.

There are various cameras and periscopes in the bath for detailed observation of activities. Should there be any problem in the top roll machines, the machines facing one another are lifted up by the action of electronic control.

## 6.3 Canal-Spout, Pouring of Glass in the Bath

The glass completes its travel to the tin bath by passing through the canal followed by the flowing over the lip stone under controlled manner. As mentioned before, the lip stone and the spout are the most critical parts of float technology. The quality of glass in the good and the bad sense is related to this part. There are of course various defects which may occur after this point (tin specks, drips, etc).

To maintain a continuous overview on glass quality, some dimensional parameters have to be checked regularly. For example, the vertical distance between the tip of the lip stone and the molten tin should be within 60–70 mm, and the distance between the spout lip and the back of the wetback tile (pull back distance) should always be 203 mm. Otherwise quality problems are likely to occur. These dimensional values should be kept stable all through the forming activity of the bath. If there is any negative development with the quality, this part is the first to attend and question. For explanatory notes see Figure 37.

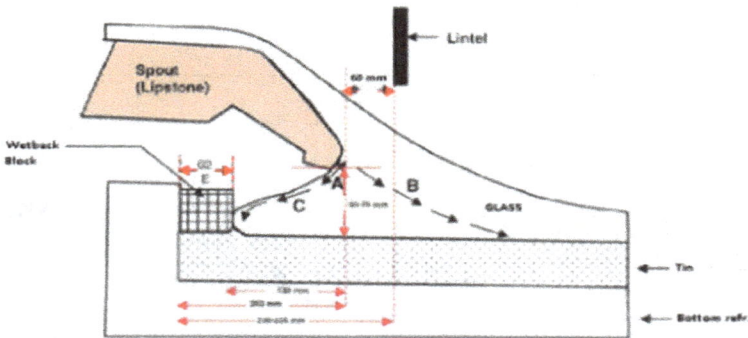

Figure 37: The details of glass flow into the tin bath. The dimensions of lip stone and wetback are most critical in float glass production.
The flow path of glass is very important to understand for the interpretation of related problems.

Only 2% of the glass poured in the bath goes back to the wetback tile before joining the normal flow. There are no direct or indirect ways of measuring this region. If the wetback region is open oxygen will enter the tin bath. However, there is an indirect method which was noted during process research. There are some features reflected onto the glass which can give an idea about the areas mentioned. The two important features are known as **wetback line** and **top surface feature**, Figure 38.

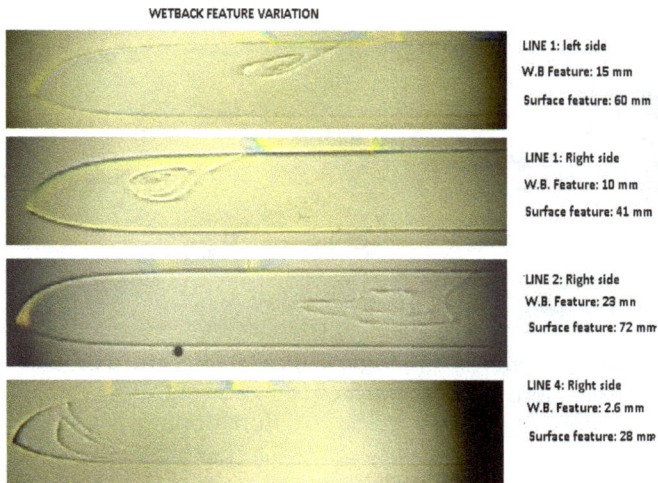

WETBACK FEATURE VARIATION

LINE 1: left side
W.B Feature: 15 mm
Surface feature: 60 mm

LINE 1: Right side
W.B. Feature: 10 mm
Surface feature: 41 mm

LINE 2: Right side
W.B. Feature: 23 mn
Surface feature: 72 mm

LINE 4: Right side
W.B. Feature: 2.6 mm
Surface feature: 28 mm

Figure 38: There are two important features which can be seen near the edges of the ribbon; **wetback feature** and **surface feature**. The curly appearances are wetback features. The starting point from the top surface to the edge is the measure of wetback (<25 mm). The top surface feature (<75 mm) appears as a set of parallel lines on the top surface (not seen in the photo). The starting values decrease in time depending on the erosion of spout and any changes in dimensions in the wetback region.

## 6.4 Investigation of Wetback Region

The area between the wetback tile and the restricter tile is called the wetback region. If the distance between the lip stone and the wetback tile is correct (130 mm) then the glass follows trajectory **C**. If this distance is narrow the glass at the tip of spout, region **A**, shows some oscillation which causes seeds. In this case the seeds follow the path that goes to the bottom of the glass, path **B**, Figure 38. The behaviour of glass currents in the wetback region are critical. Under normal conditions currents flow to the sides (right and left) whereby the defects are pushed to the sides and the bottom surface. Glass currents have a fascinating action in this region. They start from the middle, and from top to bottom in a corkscrew manner, progress to both sides which results in positioning the defects at the edges (boards).

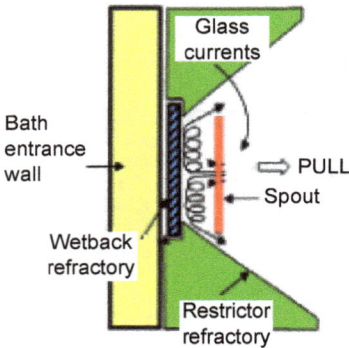

Figure 39: Glass flow currents in the wetback area. Note corkscrew-like flow.

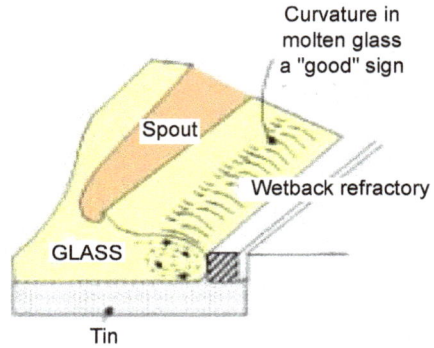

Figure 40: The curvature on glass surface is a good sign for wetback functioning.

The helical action of the glass currents forms a visible curvature of the glass. This is a good indicator of glass currents being in the right state, Figure 39. In fact, this movement is only reflected in the 2% of glass that goes towards the wetback and to the sides from there. The top surface feature is the contact trace of the glass to the spout side blocks (jambs).

To observe the wetback line and the top surface feature a glass strip of 1–2 cm, including the edge is etched in 8–10% HF solution for a period of 10 min. Then make the following measurements, Figure 41.

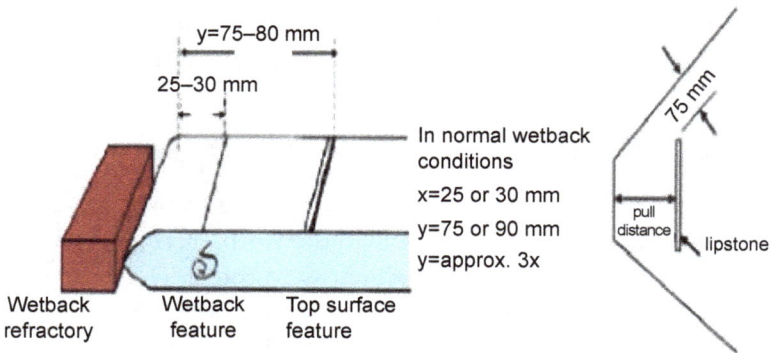

Figure 41: The measurements of wetback and surface features. The distance of surface feature is approximately three times more.

The top surface feature is due to the contact line of glass and sillimanite or high alumina jambs. The nearer this feature is to the edge of the glass, the more corrosion of jambs. In this case the pull distance is reduced. To reduce the corrosion of jambs it is advisable to reduce the temperature. When the canal temperature is reduced, the wetback line moves towards the edges. In case the wetback line disappears, it may be a choice to remove the wetback tile. There have been successful operations reported in this manner.

Under normal conditions wollastonite defect appears on the top surface, about 2 cm from the edge. The defects which are initially at the bottom surface, due to the localized currents end up at the top surface. On the other hand, if the wetback function is insufficient, the wollastonite defects show up at the bottom surface and in the middle of ribbon, Figure 42.

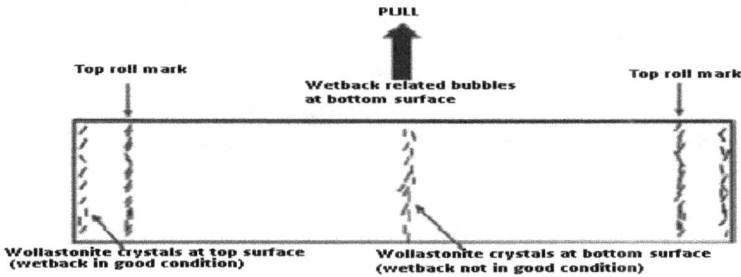

Figure 42: Wollastonite crystals on the top surface near the edge, beyond the top roll mark (if the wetback is functioning well), otherwise crystals appear in the middle at the bottom surface (if the wetback is not functioning as expected).

## 6.5 General Features of Float Glass

Almost all plain flat glass produced in the world is manufactured by float glass technology. The exclusive leader in this field is Pilkington Technology. In the early stages some attempts at differing technology has been tried (the production method of direct delivery by PPG of USA). However currently the major companies in the world (St Gobain, Asahi, Guardian, Saratov, and many companies in China) are all operating with the Pilkington Licence.

The 'defect free nature' of float glasses produced by different companies varies to some extent depending on the expertise of each producer. However, all float glasses have some properties which are common to all:

- Float glasses can be produced in thicknesses varying between 0·6–25 mm. Both thin and thick glasses need specialist knowledge, therefore some producers have an advantage at these ends. The production of thin float requires not only expertise but also a highly specialized line. Both thick and thin glasses are produced over short campaign periods.

- The levels of some oxides in float glass compositions may be flexible. There are no strictly limiting levels. It is only essential to meet the requirements of chemical and physical properties of glass for the intended use. The quality criteria for architectural, automotive, mirror, photovoltaic glass types have to be well satisfied. The competition in the market unavoidably has to take into account the price of the product, which is another pressure factor. The quality requirements on float glass may vary from country to country. This point also encourages the producer to specialize to meet customer demands.

- A common feature to all float glasses is that they are produced by floating glass on molten tin. The contact of the bottom surface with the tin results in unavoidable diffusion of Sn ions into the glass, the extent of which depends on the thickness of ribbon and also on the design. Tin content varies between 2–4%. It may be high at the beginning of campaign for a few weeks, then stabilize to a lower value. A level of 0·1% is quoted for $SnO_2$ for the top surface, but this is not normally included in float glass composition. The level of $SnO_2$ and the way it is distributed on the bottom surface can affect the surface properties. This point will be dealt with in more detail later.

- The typical range of oxides in float glass chemistry may differ depending on the source, Table 18.

  The main variation of oxides can be seen with MgO and CaO. The melting temperatures can be affected, for example replacing 1·3% MgO for 1% CaO reduces the melting point by 20°C. The reduction of the MgO/CaO ratio can improve batch free time, therefore pull can increase. The lowering of the melting temperature reduces refractory corrosion rate and the furnace emissions.

| Float Glass Consortium, 2012 | | Literature (Smerck, 2010) | |
|---|---|---|---|
| $SiO_2$ | 71·5–73 | $SiO_2$ | 72·02±0·42 |
| $Al_2O_3$ | 0–1·3 | $Al_2O_3$ | 1·00±0·13 |
| $Na_2O$ | 13–15 | $\Sigma RO$ | 12·12±0·79 |
| $K_2O$ | 0–0·7 | $\Sigma R_2O$ | 14·32±0·63 |
| $CaO$ | 7·5–10·5 | | |
| $MgO$ | 3·5–4·5 | $\Sigma R_2O+\Sigma RO=26·44±0·61$ | |
| $Fe_2O_3$ | 0·02–2·1 | | |
| $SO_3$ | 0·2 | | |

Table 18: The range of oxides in float glasses according to two sources.

- An important feature of float compositions is that the $Al_2O_3$ content of the glass needs to be carefully adjusted, so that it withstands all risks for chemical resistance. From investigation of the literature and XRF analyses of well over hundred float compositions, it was clearly established that the following range exists for $Al_2O_3$:

    75% of float glasses contain     0·5–1·5% $Al_2O_3$
    12% of float glasses contain     >1·5% $Al_2O_3$

80% of the producers in the Far East, and 20% of those in the west use >1·5% $Al_2O_3$. In general producers in the USA prefer low $Al_2O_3$, whereas producers in the Far East, due to climatic conditions, may use 2% or even up to 3% $Al_2O_3$. The numerous XRF analyses made provided an opportunity to establish a relationship between $Al_2O_3$ and alkali level, Figure 43.

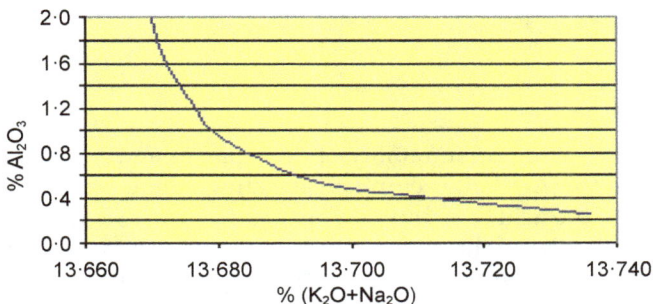

Figure 43: The relationship between $Al_2O_3$ and alkali level. If you fix alumina or alkali, one can determine the other.

It is claimed that with this relationship a more refined value for $Al_2O_3$ can be determined. It was clearly observed that as the alkali content is increased the level of $Al_2O_3$ decreases. In calculations made for the chemical durability of float glass, it was asserted that $Al_2O_3$ and $K_2O$ together affect the durability positively, whereas $Na_2O$ has negative effect, Figure 44. The points mentioned here are general aspects for float glass composition.

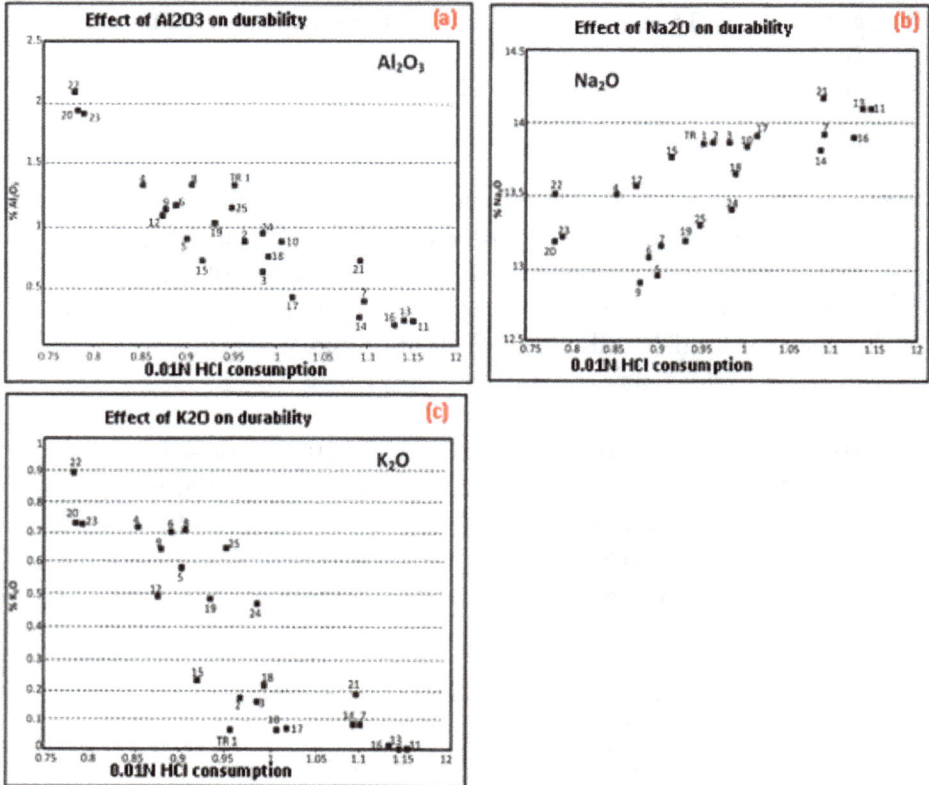

Figure 44: The chemical durability of float glasses and its relationship to $Na_2O$, $K_2O$ and $Al_2O_3$. It is emphatically shown here that as $Al_2O_3$ and $K_2O$ increase the acid consumption needed for neutralisation decreases. On the other hand the increase of $Na_2O$ causes an increase in acid consumption. In summary, $Al_2O_3$ and $K_2O$ together improve the chemical durability, while $Na_2O$ decreases durability.

- Float glass gains several properties during the production, for example the diffusion of $Sn^{2+}$ into glass, results in the culmination of the following points [56–58]:

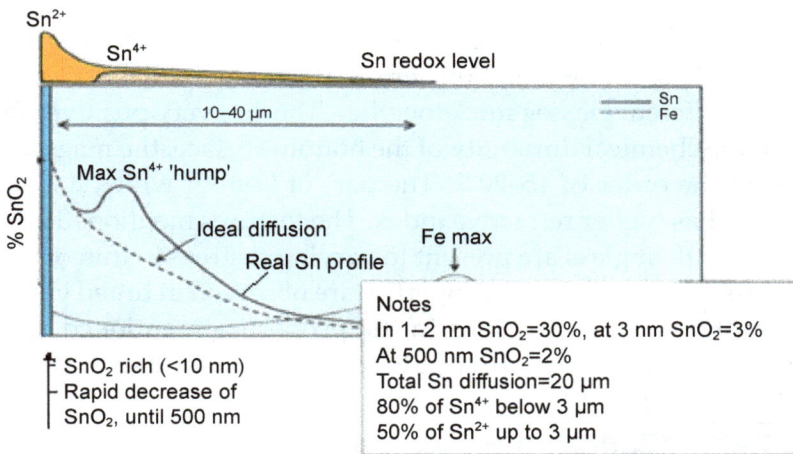

Figure 45: The characteristics of tin diffusion into glass is very important for understanding float glass. The bottom surface is therefore special. The presence of tin in the bottom surface of the glass, about a depth of 20 μm, behaves as if it is a different glass, two glasses stuck together. It is this thickness of glass that alters the properties (chemical durability of the bottom surface is 20% better, bloom, refractive index etc).

- The $SnO_2$ percentage in the first couple of nanometres may reach 30%, in the next 3 nm the $SnO_2$ sharply drops to 3%,
- In the 0·5 μm depth the $SnO_2$ reaches a value of 2%. This is why the $SiO_2$ value in the bottom surface is 70%, and 72% on the average.
- The level of $SnO_2$ in the bottom surface is about 3–4% in the first few weeks, which later levels off. $SnO_2$ is present in the bottom surface of all float glasses. The level of $SnO_2$ and the profile it shows is related to the bath it is produced in and also to the contaminants the molten tin contains. The following features are common points;
- $SnO_2$ diffusion reaches a depth of 15–20 μm. In the first 8–10 μm there is a peak (hump) in all float glasses. The occurrence of this peak is explained with the presence of $Sn^{2+}$ and $Sn^{4+}$ ions.
- 80% of $Sn^{4+}$ ions are below a depth of 3 μm, whereas only 50% of $Sn^{2+}$ are below a depth of 3 μm.
- The diffusion characteristics of other elements differ from $SnO_2$. The variations noted in the bottom surface are never seen on the atmosphere side. The variations seen in the bottom surface directly affect the physical and chemical properties.
- The first few angstroms (1 μm=1000 nm=10,000 Å) of the bottom

surface has a very different chemical composition. One can draw an analogy for the situation here: it can be assumed that there are two different glasses stuck together. This has very positive effect on the chemical durability of the bottom surface, the magnitude is of the order of 15–20%. The part of bottom which has high $SnO_2$ has higher refractive index. The features mentioned above for the float glass are present to varying degrees in thin or thick glasses. In fact, similar properties are observed in tinted glasses.

The presence of variations in the properties mentioned above may show itself in the form of **bloom** during secondary processes of glass [59], Figure 46.

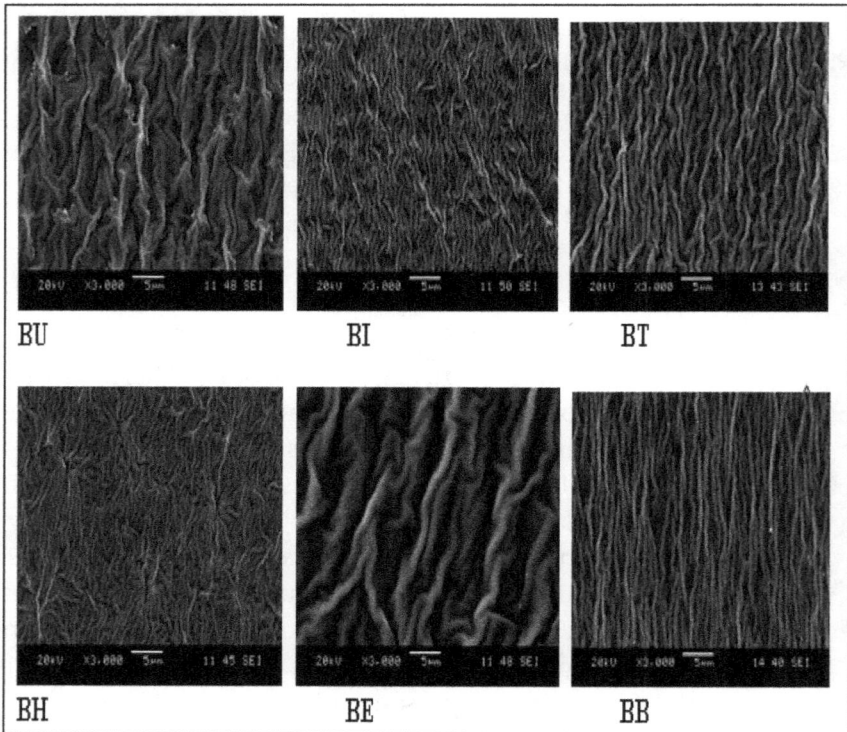

Figure 46: **Bloom** is an important defect in float glass that appears during secondary processes. The above samples are all 4 mm in thickness and are produced by different producers (the age of baths are not known). The samples were subjected to bloom test at 730°C, and were investigated under an electron microprobe (with all analytical parameters the same). There are tremendous differences, although I have some ideas but not enough to give a comprehensive explanation of the development!

If necessary precautions are not taken during the secondary processes there may be a major loss in glass. The bloom phenomenon develops at 730°C on the bottom surface. Investigations carried out on clear 4 mm thick different float glasses; there are very clear variations depending on the tin bath where the glasses are produced. The appearance of the bottom surface is very much like that of a aged skin. If the bottom surface forms the outer curvature then there will be no problem (the bottom surface is put in tension). The micro details of the surface cannot be seen by the naked eye. It resembles the appearance of heat treated metal (like haze) or tempered glass. These glasses can easily be seen during sunny days in the rear glass of cars. With Polaroid sunglasses the effect is seen more easily.

### 6.5.1 The Amount of Tin in Float Glass

$SnO_2$ present in the bottom surface affects many properties of float glass. For instance, the positive contribution on chemical durability; bloom is a source problems during secondary processing. The level of $SnO_2$ can be determined by using XRF. It is recommended to compare the $SnO_2$ with that of a good quality float glass as a standard. Through many years of Sn-count analyses carried out on float glass bottom surface, it was established that the counts are rather linear above 3 mm thickness. There is some deviation below this thickness, Figure 47 and Table 19. The reason for this is believed to be due to higher speeds of the ribbon below 3 mm (there is less contact time between molten tin and glass). The scatter of tin counts of thick glass is noticeable.

| Thickness (mm) | Amount of $SnO_2$ (g/m$^2$) |
|---|---|
| 1·6 | 0·15 |
| 2·2 | 0·21 |
| 2·5 | 0·24 |
| 3·0 | 0·29 |
| 4·0 | 0·38 |
| 5·0 | 0·48 |
| 6·0 | 0·58 |
| 8·0 | 0·77 |
| 10 | 0·97 |
| 12 | 1·17 |

Table 19: Different thicknesses of float glasses and the amount of tin in unit area (g/m$^2$) are indicated. The amount for 4 mm thick glass has been measured using electron microprobe, the other thicknesses have been calculated.

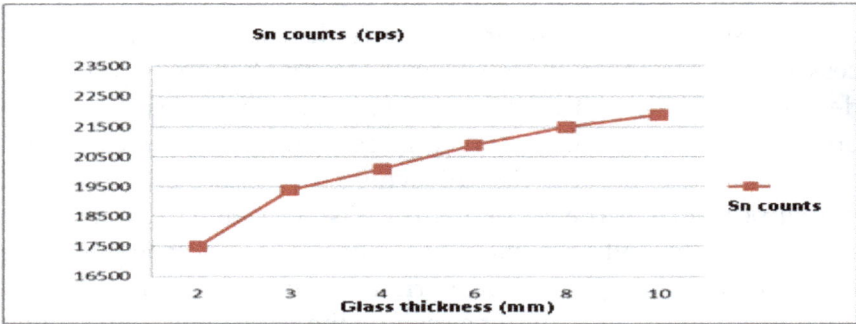

Figure 47: The relationship between tin counts (cps) and thickness of glass. Above 3 mm the line is fairly linear, but deviates below this thickness.

During the production of float glass there is a continuous loss of metallic tin (Sn), due to Sn diffusion into the glass. The amount of metal tin lost in this way is estimated to be 5–6 ton/year. Thicker glasses consume more tin, because of the longer residence time between molten tin and glass.

Normally, the loss of tin is not calculated as $g/m^2$ glass. A loss of 0·5–1·0 mm molten tin may create serious quality problems for glass; lapping bubble is one example.

Numerous electron microprobe analyses of the bottom surface has led us to believe that the $SnO_2$ content there is between 3–4%. Almost all $SnO_2$ is accommodated in the first 20 μm. The quantity of $SnO_2$ ($g/m^2$ glass) is calculated for 4 mm thick glass. This value is taken as the basis of reference, and all the values for each thickness have been calculated, Figure 48.

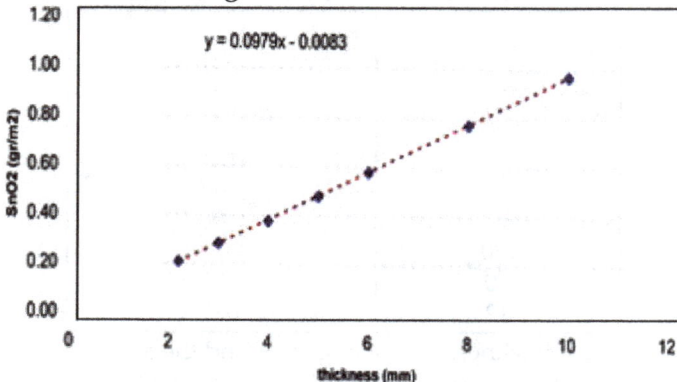

Figure 48: A linear relationship between glass thickness and tin oxide concentration.

A strong linear relationship has been confirmed with the following equation.

$Y=0 \cdot 0979X-0 \cdot 0083$

Where $Y$ is the value of $SnO_2$ (g) in 1 m² glass; and $X$ is the thickness of glass. Since the amount of glass produced is known, using the thickness values of glass, the $SnO_2$ consumed can be easily calculated.

With the experience gained in many years, it was noticed that the amount of $SnO_2$ calculated and the tin blocks charged by the operators in the tin bath were in very close agreement. Therefore, it is quite easy to calculate the amount of metal Sn to be charged; and it should be safer to do tin management and planning.

### 6.5.2 The Diffusion Characteristics of Other Oxides to the Bottom Surface

As glass proceeds from hot end (1000°C) to the cold end (600°C) there are various thermochemical reactions that take place between molten tin and glass, Figure 49.

The most important of these reactions is, of course, the diffusion of Sn into glass. As Sn diffuses into the glass, a balance is established by the diffusion of other oxides.

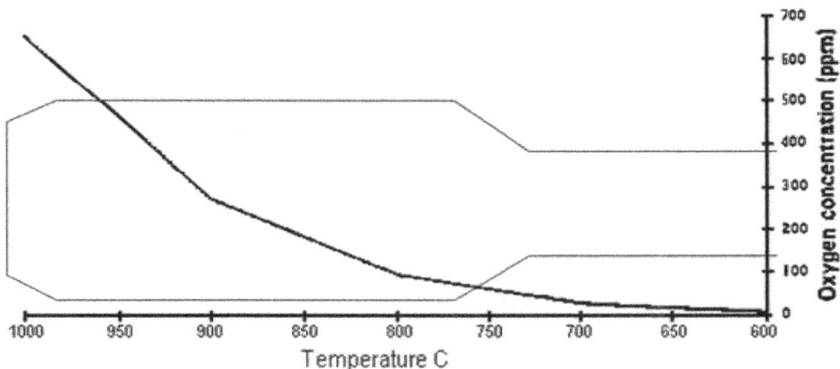

Figure 49: The solubility of oxygen in molten tin, which is maximum at 1000°C, with a concentration of 650 ppm oxygen.

The Sn profiles of glasses produced in different tin baths may differ considerably. The tin bath conditions and the thickness of glass are important parameters which affect the profiles. The profiles of glass making oxides observed in the bottom surface are displayed in Figure 50.

In general, the activity of many of the oxides tend to end at a depth of 10 μm. The profiles of tin and iron can extend to depths of 20 μm and 30 μm, respectively. The behaviour of oxides depends on the contamination load of molten tin. Oxygen is the most critical contaminant that affects the quality of glass.

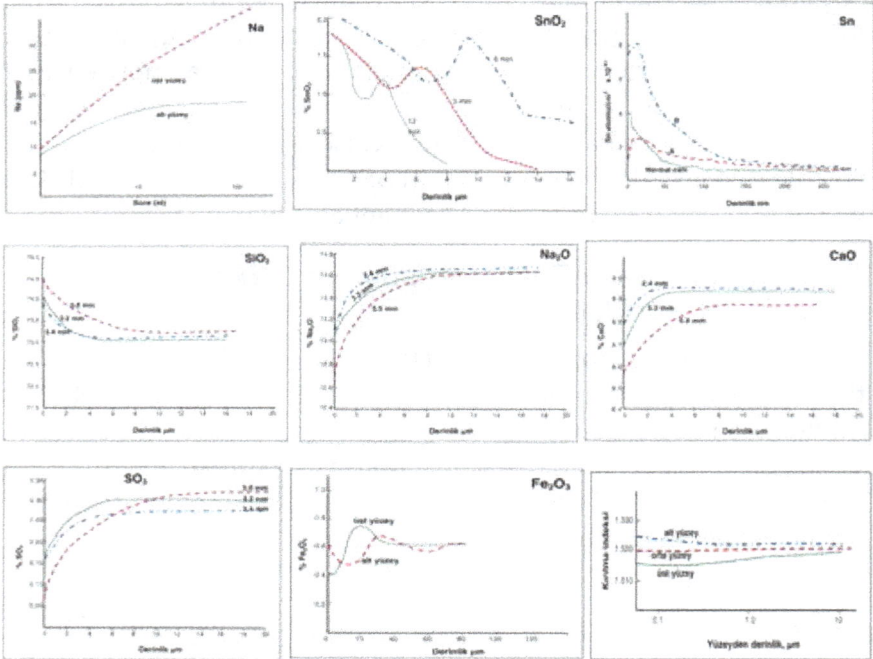

Figure 50: The diffusion characteristics of some glass making oxides ($SiO_2$, $Na_2O$, $CaO$, $SO_3$) diffuse into the bottom of glass about 6–8 μm, while $SnO_2$ and $Fe_2O_3$ diffuse approximately 20 and 30 μm. Beyond these values the base glass composition approached.

Oxygen can enter the bath in many ways, the glass itself is a continuous source of oxygen. The $N_2$ and the $H_2$ used in the bath have always some oxygen as a contaminant. These should continuously be analysed by instrumental techniques. There may be leakages of air into tin bath in several ways. The team in charge of tin bath operation should be at alert all the time, for the prevention of air leakages

No matter what precautions are taken the cumulative totals of contaminants in the molten tin increases during the campaign, Figure 51. This is known as the **contamination level of tin**, the higher the level is the more quality problems can occur in the bottom surface. An attempt should be made to reduce the contaminants by some extractive metallurgy approach in a continuous manner. If possible this type of facility should be integrated into the bath. This can be a good example of industry–university project sponsored by the glass producing company. Elements like Cu, Zn, Ag, Bi, etc, which can form an amalgam with tin may not be very critical. Special precautions must be taken for iron when tinted glass is to be produced [57,60].

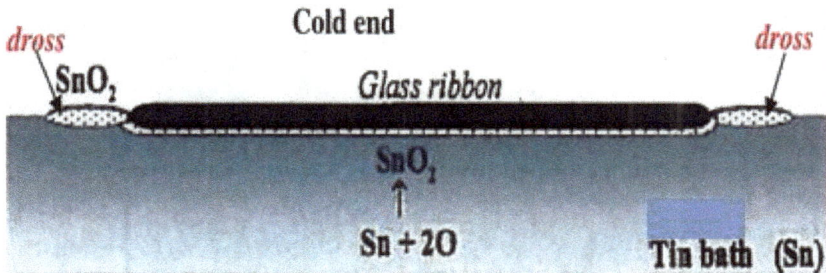

Figure 51: Glass floats on molten tin, there is a continuous reaction between the two in the form of diffusion. The elemental contamination of molten tin goes on all through the campaign.

Another important point is the reaction of oxygen with Ca and Na which can be present in molten tin to form what is called as **scum**. The reaction developed in the bottom surface are shown in Figure 52.

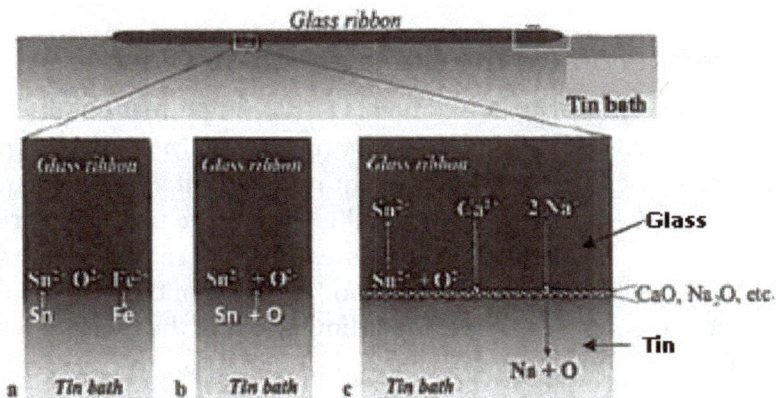

Figure 52: At tin/glass contact there is a prominent reaction of Ca, Na and O, which is more effective during coloured glass production. This is known as scum formation.

Elements which can enter the glass via the bottom surface are distributed homogenously with the help of Sn-currents. This finding is very important for logical or meaningful analytical requirements.

Samples of molten tin taken from different bays will give the same results. Therefore, as these analyses are tedious and time consuming, it should be easier to reach the results in one attempt. In other words, one sample from any bay should be enough for analyses of oxides in tin, Figure 53.

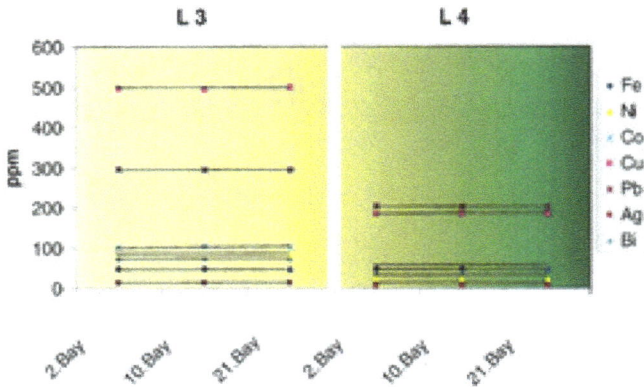

Figure 53: It is very interesting to note that the elements which are present in molten tin have the same values and do not change from bay to bay. This investigation was made for two different baths, L3 is older L4 is a new bath.

### 6.5.3 Defects on the Top Surface Float of Glass

The most striking property of float glass is its perfect quality. In a time where customers are continually narrowing the limits for quality. Some glasses are traditionally required to be of the highest quality; i.e. automotive glass, mirror and glasses to be used for various coating purposes. Architectural glasses have more relaxed quality limits. The quality of thin glass is another point of challenge. Some bottlenecks on quality may be due to limitations of the tin bath, which are critical for optical distortions of various types.

The most widespread defects seen on the top surface originate from the bath atmosphere effect on the roof and those which originate from a variety of coolers, Figure 54.

## Top Dirt

All defects result from contamination of the tin by sulphur and oxygen. Tin sulphide and tin oxide can volatilise into the bath atmosphere and eventually condense on cold areas such as coolers, periscopes, top rolls, fences, pushers, etc. and the bath roof. Defects occur when these condensates fall onto the ribbon.

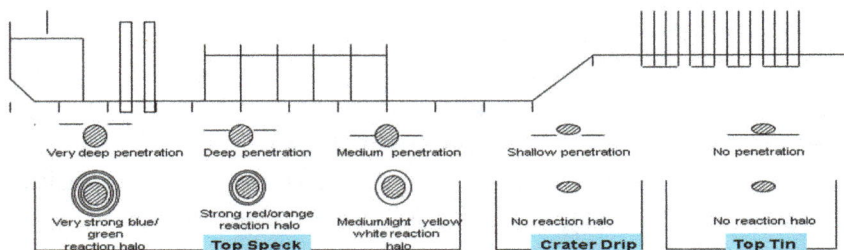

**1. Reduction.** If the tin sulphide and/or tin oxide are reduced by hydrogen (promoted by heat, e.g. reheat, removal of coolers etc.) then particles of elemental tin will fall onto the ribbon.
**2. Physical. E.g.** If coolers are banged tin sulphide and/or tin oxide may fall on the ribbon.

Figure 54: All defects arise from the contamination of tin by oxygen and sulphur which volatalise into the atmosphere. These condense on cool places, coolers, to form what is called as top speck, crater drips and top tin.

The defects tend to increase with the campaign. The main driving mechanism is the increase of dissolved oxygen and sulphur in the molten tin. Oxygen and sulphur tend to evaporate in the form of SnO and SnS at the hot end. The evaporation of glass at the tin bath contributes to the increase of SnO and SnS. Oxygen, as mentioned before, can find its way to the bath by air leakages. For sulphur the following reactions develop;

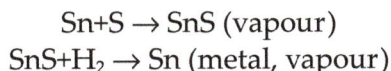

$$Sn+S \rightarrow SnS \text{ (vapour)}$$
$$SnS+H_2 \rightarrow Sn \text{ (metal, vapour)}$$

Oxygen encourages the SnO evaporation, which reacts with $H_2$ to produce Sn vapour; this precipitates on the roof and finally falls on the surface of glass in the form of top specks (metallic tin spheres). The water coolers are potential areas where this precipitation takes place. The reactions mentioned are given below;

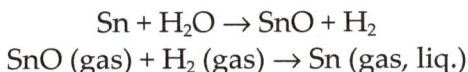

$$Sn + H_2O \rightarrow SnO + H_2$$
$$SnO \text{ (gas)} + H_2 \text{ (gas)} \rightarrow Sn \text{ (gas, liq.)}$$

## 6.5.4 Formation of Tin Specks and Investigation Methods

Tin specks are found on the top surface of float glass. As glass enters the tin bath sulphur in the composition begins to evaporate. Sulphur dissolves in molten tin easily in the hot end. The next stage is the evaporation of SnS. This reaction which is temperature dependent is a continuous one. As the pull of the furnace increases (more glass to the bath) the amount of sulphur transferred to the bath also increase.

As the temperature falls, in other words towards the cooler parts, these vapours find a chance to precipitate on coolers. The deposition temperature of SnS vapours is 550°C, these are reduced by $H_2$ which is present in the atmosphere to give metallic tin spheres (tin specks).

Tin specks originating at the hot end are very much smaller in size than those formed at the colder areas. Tin specks are usually released after a temperature increase. Tin specks can be present in the holes of the heating elements, the roof and the superstructure joints, and on all the water coolers.

The most critical place for tin specks is at the beginning of narrow end (57%) as indicated in Figure 55.

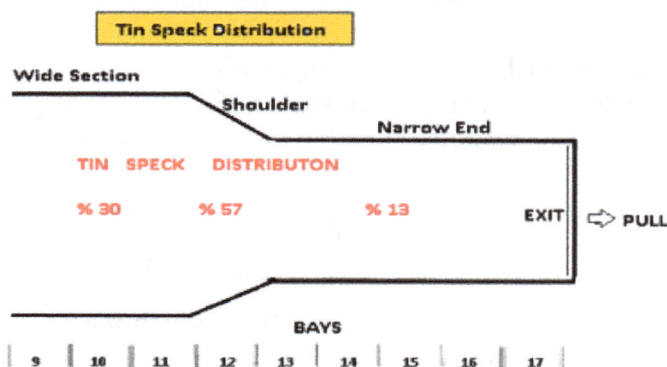

Figure 55: Maximum condensation of bath atmosphere occurs around the shoulder, where 57% of the defects happen.

As a general rule it is safe to say that the tin specks start to come off or disturb the glass quality after one or two years of the campaign. During this time the pores of the roof refractories absorb SnS and after a certain level a reducing reaction with $H_2$ gives rise to tin specks.

The percentage of $H_2$ and $H_2/H_2O$ are important factors in tin speck formation. The increase of oxygen does not cause any increase in tin specks; however bottom surface defects can occur. If $H_2$ concentration in the atmosphere is too much, tin speck concentration may increase. In the early times of this technology the % of $H_2$ used was 8–9%; which is quite high. In general, 6% is used currently. The minimum level of $H_2$ was found to be 2%, but Pilkington preferred to use 4–6% $H_2$. There are two possible reasons to reduce the level of $H_2$; one is to reduce the tin specks, and the other is to cut the cost of $H_2$. Even at low $H_2$ levels the tin bath sealing has always been important. The thickness of glass has very small effect on temperature distribution, Figure 56.

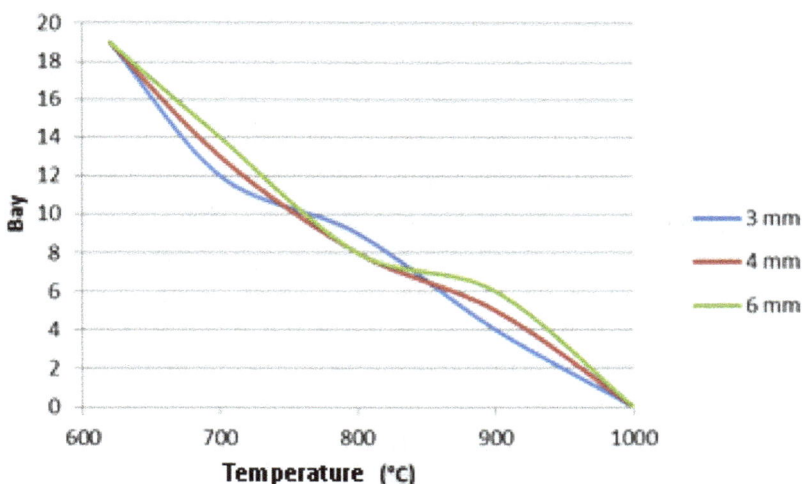

Figure 56: The temperature profiles for different thicknesses of glass do not change very much.

### 6.5.5 Tin Specks and Glass Quality

Concerning the quality of glass, there are three different sizes of tin specks, for which limits are imposed, Table 20:

| Tin speck type | Tin speck size (mm) | Limits allowed for 320×100 cm of mirror glass (number) |
|---|---|---|
| Prominent top speck (coarse size) | >0·1 | 0 |
| Top speck | 0·06–0·09 | Max. 10 |
| Fine top speck | <0·06 | Max. 25 |

Table 20: The number of tin specks and their size range are given. These limits are closely followed by customers.

Tin specks should not be concentrated in a certain area, Figure 56. The trace left behind by a speck on the top surface is evaluated as an open bubble.

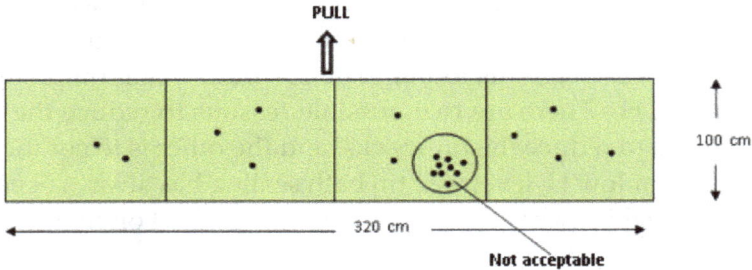

Figure 57: Tin specks are not allowed to be concentrated in any part.

## 6.5.6 Investigation Method (Diameter of Speck and Crater Depth)

Tin specks penetrate to a greater depth at the hot end because of lower viscosity of glass. Observations made by Pilkington researchers have shown that above 850°C temperatures, there is an empirical relationship between the diameter ($d$) of speck and the crater depth ($h$) that is $h=1/3d$, Figure 58.

It is quite easy to measure the values of $d$ and $h$ by using a petrographic or polarisation microscope. Below 750°C, as the viscosity is increased, the tin specks do not submerge into the glass. The crater depth and temperature relationship are indicated in Table 21.

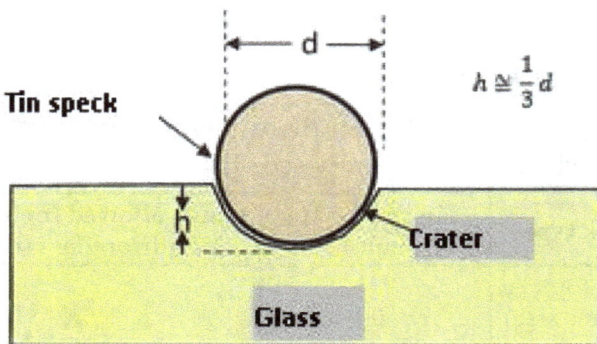

Figure 58: The relationship between tin speck diameter and crater depth.

| Crater depth ( μm) | Glass temperature (°C) |
|---|---|
| <2 | <700 |
| 2–5 | 730–755 |
| 5–10 | 750–780 |
| 10–20 | 775–800 |
| >20 | >800 |

Table 21: The crater depths and corresponding glass temperatures.

## 6.5.7 Investigation Method (Temperature Calculation)

The size of the crater caused by the tin speck depends on the diameter of the speck and the temperature of the glass, i.e. viscosity of glass, Figure 59.

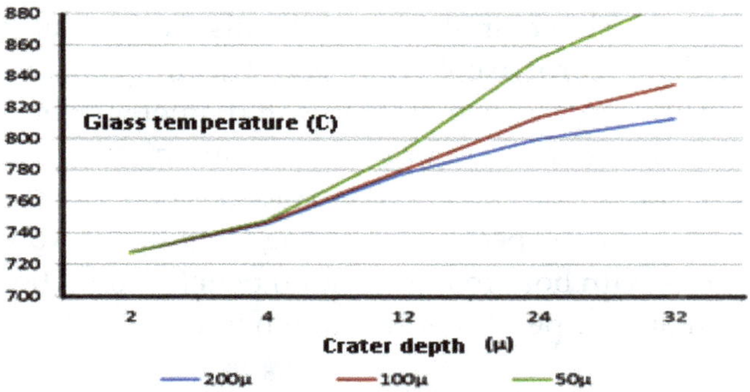

Figure 59: Glass temperatures and crater depths for different size tin speck.

| Size of tin speck (μm) | Temperature (°C) |
|---|---|
| 50 | 800 |
| 100 | 830 |
| 100 | 800 |
| 100 | 775 |
| 100 | 755 |
| 150 | 800 |
| 200 | 800 |

Table 22: Size of tin speck, crater depths and relate glass temperatures.

## 6.5.8 Calculating Glass Temperature Using Crater Depth

From evaluations made by Pilkington the temperature of glass can be calculated using tin speck (d) size and crater depth (h). This relationship may not be valid for very small sizes.

$h_e=0.35*d*\log_n[0.35*d/(0.35*d)-h]$

    $d$: crater diameter (μm)

    $h$: crater depth (μm)

Temperature (°C)=$[5754/(11{,}046-\log_{10}*h_e*t_c)]+210$

$t_c$: Cooling speed, °C/s ( ~1.2°C/s)

### 6.5.9 Investigation Method (According to Bays)

The shape of craters are almost characteristic for each bay, the temperature distribution for a 19 bay tin bath are given for ease of reference, Figure 60. The shape of craters are diagrammatically presented, Figure 61.

The tin specks originating from bays 8–12 are nearly all the same size, and therefore have similar crater details. Tin specks from these bays always have a brown coloured halo. The tin specks, depending on the viscosity of glass, penetrate to half diameter depth.

Sizes of tin specks and craters in bay 14 are highly different, the periphery of the tin speck penetrates the glass only 5 μm in order to hold on. In general specks originating from the narrow end just sit on the glass with bottom flat, with no penetration. The bays and the approximate temperatures are shown in Figure 60.

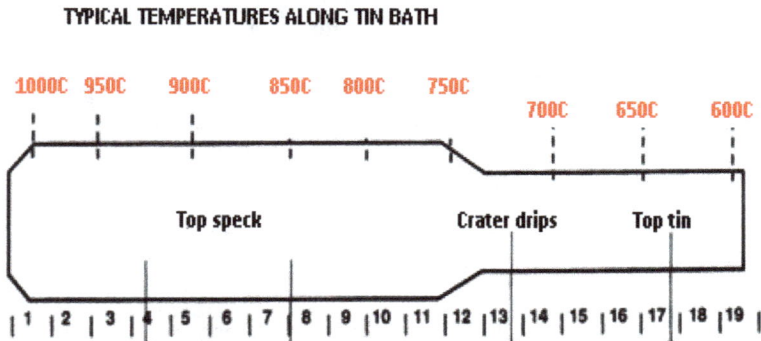

Figure 60: Typical temperature values in a 19 bay tin bath.

It is a traditional daily control action, as the ribbon is running to touch the glass where the temperature is low enough, and see if there are any tin specks in the palm of hand. From the diagrammatic presentation it is tentatively obvious that the specks are released from the narrow end, as others from the hot end will penetrate the glass and will not be released.

TIN SPECKS AND BAY RELATIONSHIP

| BAY | CRATER SHAPE | SPECK SIZE | TEMPERATURE °C |
|---|---|---|---|
| BAY: 8-12 | 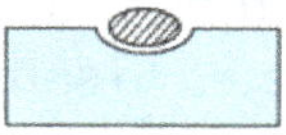 | < 95 μm | > 850°C |
| BAY: 13 |  | < 120 μm | 800-850°C |
| BAY: 14 | 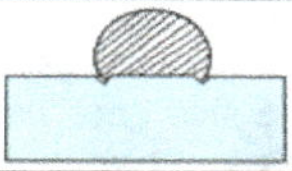 | < 700 μm | |
| BAY: 15 | 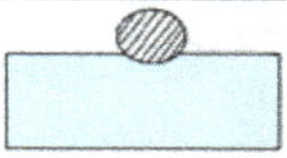 | < 200 μm | 700-750°C |
| BAY: 16 |  | < 70 μm | |

Figure 61: Tin speck sizes, their crater depth and related temperatures.

## 6.6 Tin Bath Defects

### 6.6.1 Cassiterite Defects from the Roof

Cassiterite ($SnO_2$) is an important defect which occurs on the top surface. The leakage of oxygen (air) into the tin bath from various points gives rise to the formation of cassiterite at the hot end (bay 0–1) and at some bays at the exit end of the bath. The appearance of crystals originating from these places are quite different. Those formed at the hot end are like cauliflower and penetrate into the glass due to low viscosity of glass, and those formed at the cold end are much simpler crystals. The only way to differentiate these crystals is the amount they penetrate the glass. In summary it is safe to say that those which are on the surface are from the cold end, whereas if the crystals penetrate the glass they are from the hot end.

It is possible to get some interesting textures on samples collected at the cold repair, the roof material from hot and cold end, both of which are grey in colour. In samples from the cold end it is very likely to get needle or hair like crystals of cassiterite.

## 6.6.2 Seeds and Bubbles Originating from the Tin Bath

Seeds and bubbles are always present to some extend in float glass production. Seeds which are related to the bath appear on the top and bottom surface of glass, Figure 62. Seeds do not have enough time, or the viscosity of glass does not allow them to position themselves in the body of the glass. Therefore, those bubbles or seeds which are found in the body are usually related to sources before the bath.

# Seeds and Bubbles of Tin Bath Origin

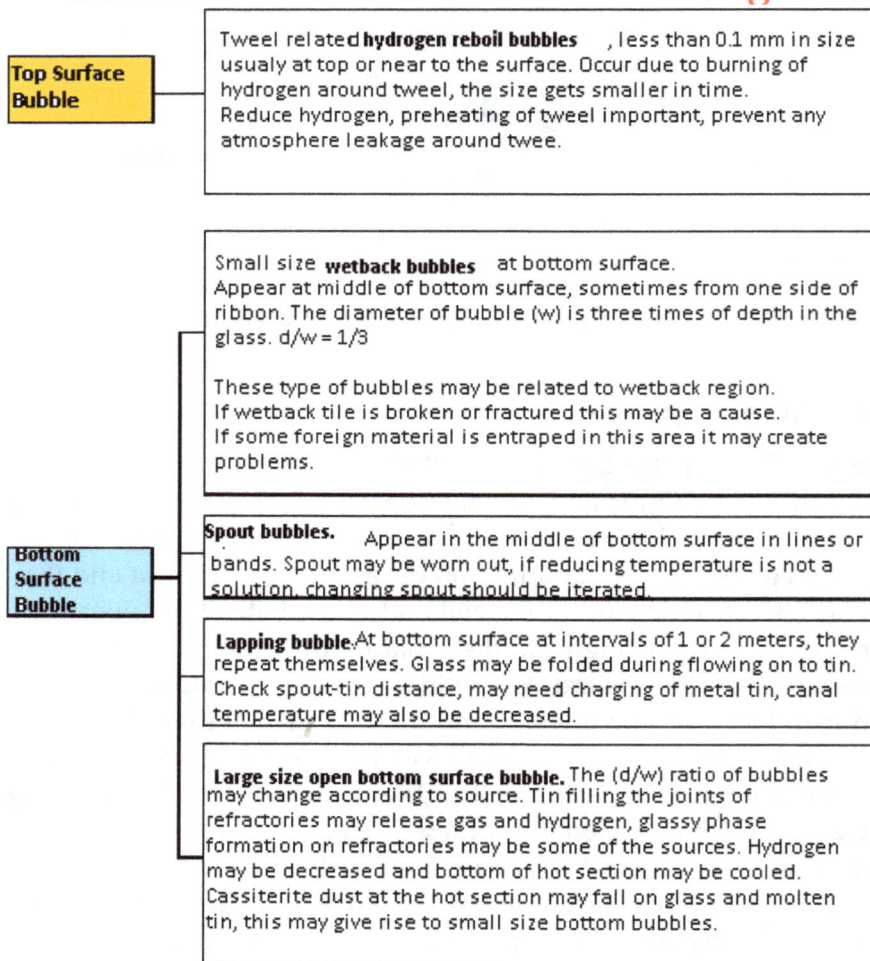

**Top Surface Bubble**

Tweel related **hydrogen reboil bubbles**, less than 0.1 mm in size usualy at top or near to the surface. Occur due to burning of hydrogen around tweel, the size gets smaller in time.
Reduce hydrogen, preheating of tweel important, prevent any atmosphere leakage around twee.

**Bottom Surface Bubble**

Small size **wetback bubbles** at bottom surface.
Appear at middle of bottom surface, sometimes from one side of ribbon. The diameter of bubble (w) is three times of depth in the glass. d/w = 1/3

These type of bubbles may be related to wetback region.
If wetback tile is broken or fractured this may be a cause.
If some foreign material is entraped in this area it may create problems.

**Spout bubbles.** Appear in the middle of bottom surface in lines or bands. Spout may be worn out, if reducing temperature is not a solution, changing spout should be iterated.

**Lapping bubble.** At bottom surface at intervals of 1 or 2 meters, they repeat themselves. Glass may be folded during flowing on to tin. Check spout-tin distance, may need charging of metal tin, canal temperature may also be decreased.

**Large size open bottom surface bubble.** The (d/w) ratio of bubbles may change according to source. Tin filling the joints of refractories may release gas and hydrogen, glassy phase formation on refractories may be some of the sources. Hydrogen may be decreased and bottom of hot section may be cooled. Cassiterite dust at the hot section may fall on glass and molten tin, this may give rise to small size bottom bubbles.

Figure 62: Seeds of tin bath origin appear on top and bottom surface of glass. The causes and precautions are listed in the above figure.

The shape and size of seeds, their intensity and distribution in the glass, if they are recurrent (lapping bubble), may be important bits of information for interpretation. The chemical analyses of bubbles provides very significant information.

In general, at the beginning of the campaign, because of the possibility of several sources, there may be a high level of seeds, the intensity of which decreases with time. The seeds from the melter and the conditioning end tend to decrease and stabilise. A similar trend can be expected for seeds originating from the bath.

The bottom refractory blocks are subjected to several diamond sawing and cutting operations; there are always some small refractory particles which lodge themselves in the open pores. During the operation these particles release themselves and are the cause of open bubbles on the bottom surface. This type of disturbance may go on for a few months. It is frequently, quite possible to see relics of refractory particles in the open bubbles. A special measure should be applied to clear off the particles, e.g. by strong suction. The refractory producers should be encouraged to undertake this cleaning.

## 6.7 Problematic Areas in Float Glass Production

Even though float glass is the best quality of glass, the presence of several problems during production are unavoidable. Some of these problems are related with the tin bath; on the other hand some of the problems are those which are related with the melter and the conditioning end. In this section more space will be devoted to those problems directly related with the tin bath. The problems are investigated under three headings [61], Figure 63:
- Top surface (atmosphere side) defects,
- Defects related with the body of glass,
- Bottom surface (tin side) defects.

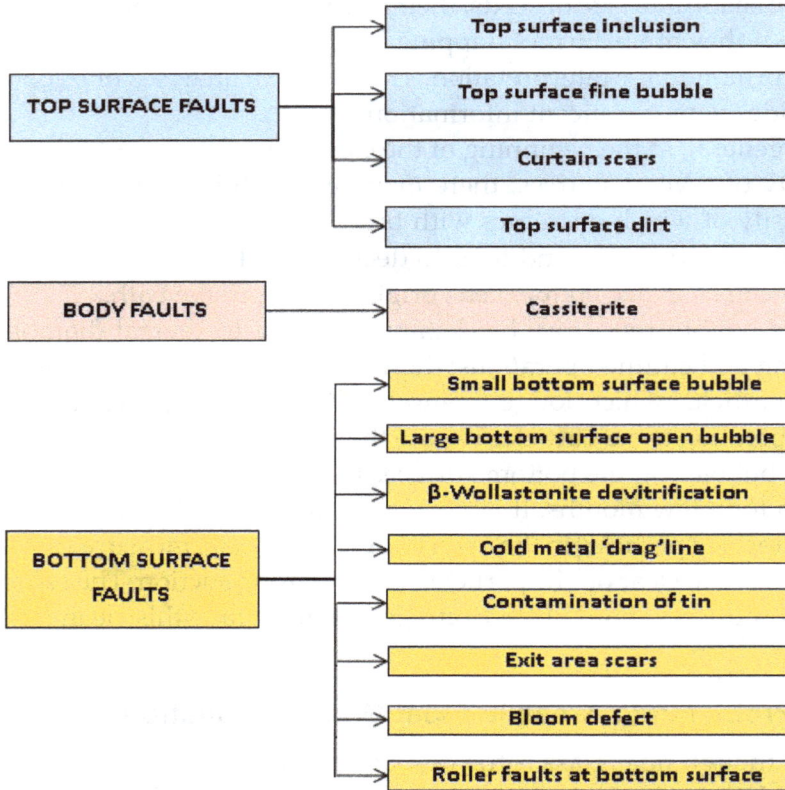

Figure 63: Bath defects have been classified under three headings
according to the literature: top and bottom surface defects and body
defects. Defects related to rollers occur outside the tin bath but the
defects are considered as bottom surface defects.

Unfortunately, since the beginning of this technology there has
not been a comprehensive literature covering the possible problem.
No producer has been open enough; preferred to remain on the
conservative side. All information was and probably still is deemed
to be secret. The information reflected here is a collection of 30 years
of experience based on meeting notes, technical round table chats
during stand ups or revaluation of loose paper note pads, various
ways of personal communications. None of the information is in
anyway considered to be strategic or even secret anymore. It is possible to have access to any topic through the internet. These are bits

and pieces of information brought together by personal endeavour, with the hope of some benefit to younger generation.

Based on the information made available by Pilkington, an approach for classification of defects has been put forward. This information is used and common to all licensees of the technology.

Comprehensive defect identification is vital for precautions to be applied. For instance, the size, shape, crater depths, halo, etc. are important clues to reading the details in depth of tin specks on the top surface which will concentrate all the effort towards the right bay. The tin specks falling on the top surface, as a result of roof cleaning, or increase of temperature provides a wealth of information. The contaminants oxygen and sulphur in molten tin are the main causes behind SnO and SnS evaporation, and also the formation of dross, Figure 64.

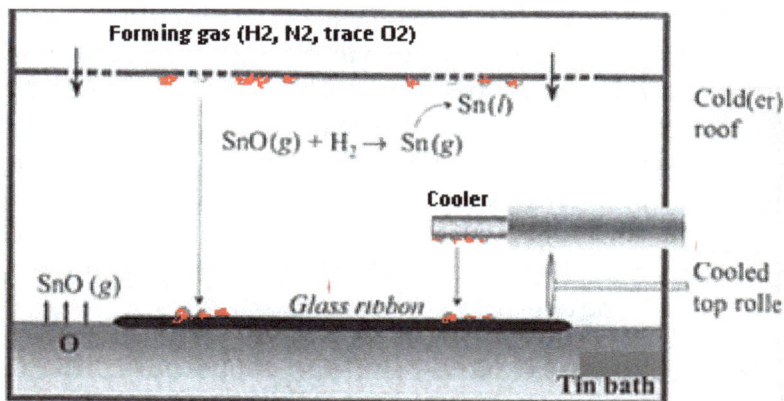

Figure 64: The formation of tin specks and dross due to evaporation of SnO and SnO from molten tin.

## 6.7.1 Devitrification Defects in Float Glass

Wollastonite which is a devitrification product is seen quite frequently in float glass. It is seen in the middle of the ribbon, on the bottom surface and in the sides (boards). Seeds and bubbles occur in the middle of the bottom surface before the devitrification defects.

According to the experience of Pilkington wollastonite is a devitrification product related to primarily to the wetback area. However, one should also take into account the possibility of canal as a source. In this case it is difficult to explain how the defects come from the middle of the ribbon. Devitrification at the canal can happen at the sides and bottom, Figure 65.

**CANAL**

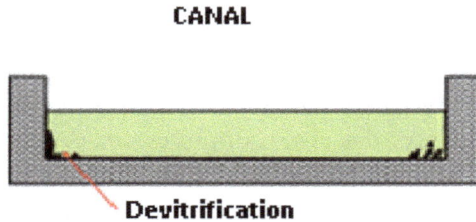

**Devitrification**

Figure 65: Devitrification defects are not rare in float glass production; one possibility is the canal, the bottom corners are possible locations.

Glass which can be on the wetback tile can also be subjected to devitrification yielding a mineral called diopside, which may find its way to the glass.

In order to prevent devitrification in the wetback area more glass should be transferred to this region. This action will result in an increase in temperature. To be able to achieve this the distance between wetback and the spout has to be increased. The temperature in the casing may cause a small increase in the bottom refractory temperature, which may also prevent the formation of devitrification. As the canal temperature is decreased then the wetback line moves towards the edges.

### 6.7.2 Bath Pressure

Bath pressure should not be below the pressure of conditioning, that is this atmosphere should not force its way to the tin bath. In other words the pressure of conditioning should be slightly lower than the bath pressure. In this way no oxygen from the conditioning end is allowed to enter the bath. The high pressure atmosphere in the canal may dislodge cassiterite from the surface of refractory.

### 6.7.3 Temperatures in Tin Bath Exit

If there is no problem with the quality of glass the temperature can be slightly high. For instance, for 3 mm thick glass 615°C exit temperature can be a little bit otherwise corrugation may occur. In general, if the tin depth is high, then the temperature difference between the middle and the sides can be small. In thick glass this temperature difference can be slightly higher, Table 23. For instance, for an increase of 20°C in the exit (615+20=635°C) is rather a high temperature.

| Tin depth (mm) | Glass thickness (mm) | Temperature difference between centre and sides (°C) |
|---|---|---|
| 55–57 | 5 | 20 |
|  | 6 | 25 |
|  | 8 | 28 |
| 83 | 3–4 | 10 |
|  | 6 | 17 |

Table 23: The glass temperatures at the exit end may vary with thickness. There may be differences of up to 28°C between the centre and the sides.

### 6.7.4 Canal Temperatures

The temperature of the canal is of critical value. The temperature is directly related with the composition and viscosity of glass. In other words, if $Na_2O$ increases the viscosity is decreased. Viscosity should be kept stable in parallel with temperature. If there are minor compositional changes the canal temperature should not be allowed to change. The glass thickness has a direct effect on temperature.

## 6.8 Surface Changes in Float Glass

Thickness variation parallel to the pull are characteristic of float glass. The terminology used for this is **Broken Line (=Ripple or short wave)**. If there are minor compositional and temperature changes due to variation in stretching this problem may arise. The coolers in the hot section may have a role here.

Rapid cooling should not be allowed for which a correct configuration of cooler is necessary. This can be ascertained by some trial and error. As pull increases this type of variation may increase.

Distortions in thick glass are seen on the top surface, whereas in thin substance they can be seen in both surfaces.

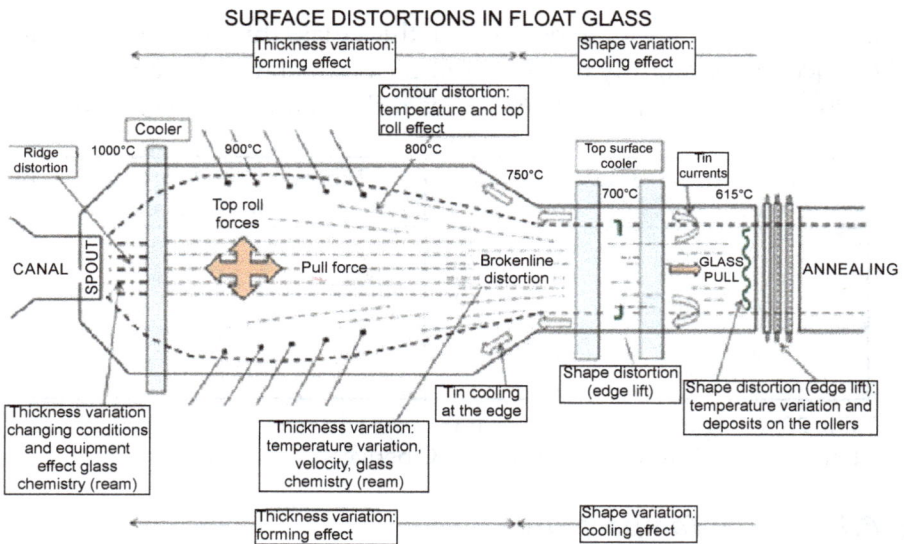

Figure 66: During the forming operations in the tin bath, due to stretching or narrowing, some shape or thickness variations may occur in the glass. Thickness variations happen in the wide section, while shape variations occur in the shoulder (broken line distortion are typical) and narrow section. Edge distortion can occur before the exit end of the narrow section. Distortion in thick glass happens on the top surface, in thin glass in both surfaces.

Shape variations are encouraged due to the tin currents in the narrow section, Figure 66. This effect is more pronounced in high pull (more tin current activity). This effect is especially critical for thin glass. Distortion of thin glass may be enhanced during lamination operations. To minimise this problem for thin glass it is recommended to stick the top surfaces of glass face to face.

Ripple distortion can be reduced by using suitable and harmonious heating and cooling. The slew angles of toproll machines can also be reduced. It is highly advantageous to manage the tin currents in case of noticeable thickness and shape variations. Multidepth tin and barriers are used exclusively to minimise this problem. There is more successful tin current control with multidepth tin.

## 6.8.1 Distortion in Float Glass

The property of float glass that stands out most for customers is its optical quality. There are some problems which affect this quality, thickness and shape variations are summarised in Figure 67.
- Thickness variation in glass,
- Surface distortions and waviness,
- Ream (chemical inhomogeneity).

| Properties | Causes and distortions |
|---|---|
| **Brokenlinple/shortwave)** | |
| 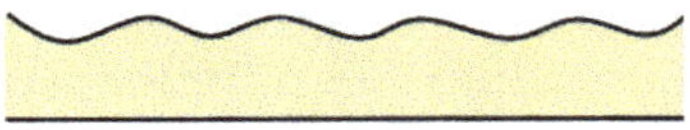 Thick glass   6 mm | Coolers at the hot end very effective. Temperature and chemical changes along the width remarkable. Top surface in thick glass more affected. It is a thickness variation. **Strong in transmission: medium in reflection** |
|  Thin glass   2 mm | Top surface in thin glass more affected, it is a thickness variation. **Strong in transmission: medium in reflection** |
| 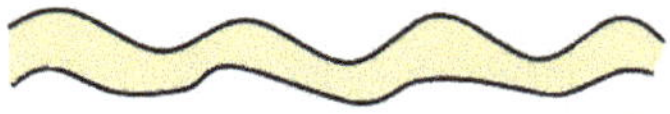 Thin glass   <2 mm | With increasing speed the effect is more pronounced. It is a shape variation. **Weak/medium in transmission; strong in reflection** |
| **Edge lift** | |
| 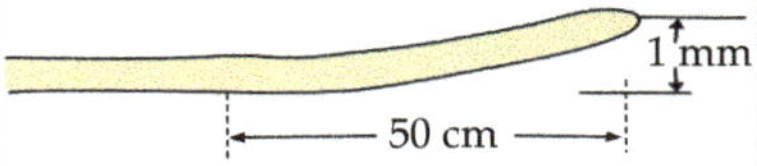 1 mm    50 cm  Excessive tin cooling | Due to tin cooling effect 50 cm of glass from the side is affected. The edge of glass lifts up 1 mm. **Weak/medium in transmission; strong in reflection** |
| 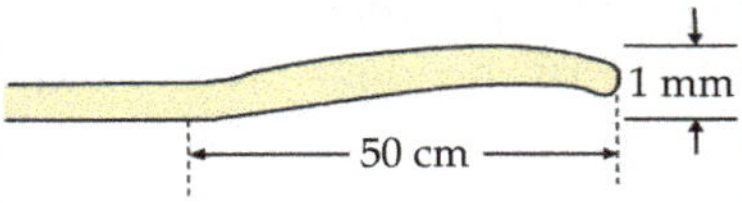 1 mm    50 cm  Excessive top surface cooling | Due to excessive top surface cooling 50 cm of glass may bent down. **Weak/medium in transmission; strong in reflection** |
| **Ream** | |
|   | The inhomogeneity level of glass is important. Distortions in the laminated nature of glass appear as optical distortions. No thickness variation. Effective stirring of glass at the waist may help. **Medium/strong in transmission; weak in reflection** |

Figure 67: Thickness and surface variations are shown. Thick glass is affected on the top surface; thin glass can be affected in both surfaces giving thickness and shape variations. Edge lift is more critical in relatively thin glasses. Ream shows up because of the distortions of lamination of glass.

Surface flatness quality is related to the production itself. As a general rule, thin glass is produced at high speed and therefore some waviness of different amplitude parallel to the pull may happen. This may appear as distortions to the end user. Some shows up in reflection mode. The amplitude is rather high; but those of ream are of narrow amplitude. The effect of thickness and that of ream are quite similar as distortion, Figure 84.

## 6.8.2 Ridge Distortion

Ridge distortion occurs due to frequent thickness changes. Abnormal operation practices, defective equipment and low quality glass can enhance ridge distortion. Bottom refractory block lamination, splitting can change the effective depth of tin, which in turn may give rise to ridge distortion. Deposits gathered behind the tweel, and the accumulation of glassy phase of different history on the tweel can give rise to ridge distortion. Distortions 30 cm from the edge may be a sign of broken tweel edges. The tweel projection with respect to the ribbon is as follows, Figure 68:

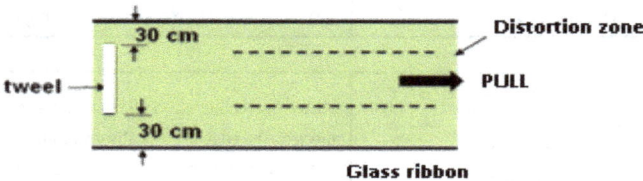

Figure 68: The projection of tweel on the ribbon. The sides (right and left) are prone to distortion, in this case check the tweel edges for any flaw.

Fractures in the lipstone and the tip of tweel may give rise to strong distortions. Any contamination behind the tweel (pieces of thermocouple, etc) are another cause for distortions. If there is any distortion within 30 cm from the edge this may imply there is a fracture at the tip of the tweel.

According to Pilkington practice glass for mirror quality is inspected more effectively in reflection rather than Zebra angles. Semi Schlieren, and shadowgraph methods can be used to control the glass quality.

**Edge lift** is caused due to upward or downward bending of glass edges. Excessive top surface cooling of glass or tin cooling may be the cause behind the edge lift problem.

Glass flow currents in the canal are unidirectional, no reverse flow here. The two elliptical appearance in glass seen at the Striagrams are glass volumes which cannot enter the canal, but return backwards from the sides. This is old glass, other parts represent new glass, Figure 69.

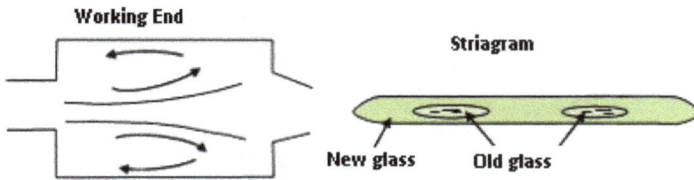

Figure 69: Striagrams are useful for showing the effect of glass currents. Old glass is situated in the body of new glass.

## 6.8.3 'Ream' as an Inhomogeneity in Float Glass

There are several inhomogeneity types in glasses, which are one or two dimensional. Cords are one dimensional and are named as striae (in float glass), catscratch (in container) and knots are also one dimensional. Ream is in the form of a layer and is reckoned to be two dimensional. Investigation methods used for cord and ream are the same. Ream is essentially different in chemical composition compared to base glass and displays a laminar structure, Figure 70, [62].

More broadly, ream can also be due to varying viscosity and temperature or it may be related to chemical changes. The effect of glass flow has a role in the final appearance of ream. In general, it is assumed to form from the mixing of glass that has evaporated in cooler and stagnant parts of the furnace, resulting in the enrichment of $SiO_2$ (+0·5% $SiO_2$) and a deficiency of alkali.

It is usually the top surface of glass which is subjected to this development. In this sense it is a mixture of two different glasses, one of which is more enriched in $SiO_2$. It is a complex mechanism and as a result it is difficult to delineate which factors have dominated. The general features of these layers are explained in the diagram.

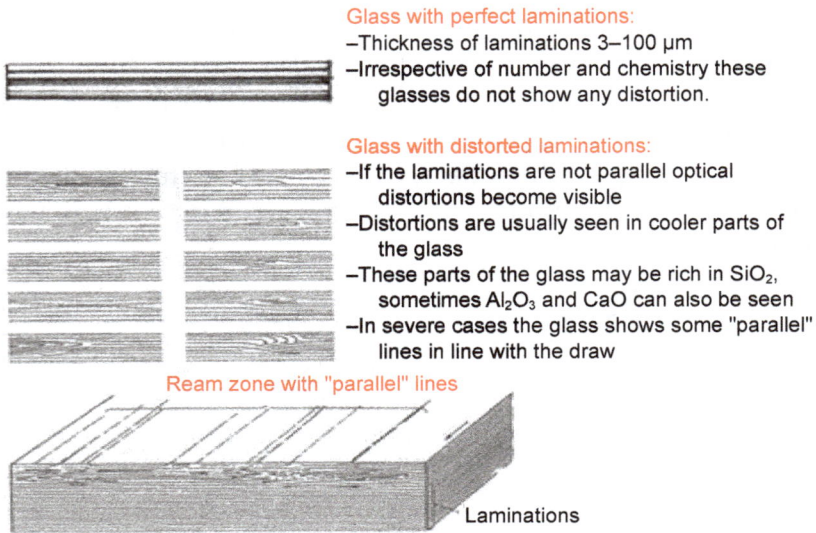

Glass with perfect laminations:
−Thickness of laminations 3–100 µm
−Irrespective of number and chemistry these
  glasses do not show any distortion.

Glass with distorted laminations:
−If the laminations are not parallel optical
  distortions become visible
−Distortions are usually seen in cooler parts of
  the glass
−These parts of the glass may be rich in $SiO_2$,
  sometimes $Al_2O_3$ and CaO can also be seen
−In severe cases the glass shows some "parallel"
  lines in line with the draw

Ream zone with "parallel" lines

Laminations

Figure 70: **Ream** is considered to be an important defect in float glass. The laminar nature of glass is disturbed by forming operations. In reality, as it is widely accepted, ream is a chemical inhomogeneity. The only variation has been determined in $SiO_2$ (0·5–1·0%).

## 6.9 Lift out Rollers (LOR) in the Exit End

Having completed the forming operations in the tin bath the glass is transferred over to the rollers followed by the annealing lehr. The glass which is in the viscoelastic state comes in to contact with metal for the first time. Glass is now in a different environment. The surface quality of LORs is of high importance for the quality of glass, as this is the first contact medium. There are three reactions which may take place here:

- **Liquid/glass contact**: Dross which can be present on the top surface of tin may adhere to the bottom surface of glass.
- **Solid/glass contact**: Material harder than glass may stick on the rollers, can cause scratching of the bottom surface, leading to surface defects of glass.
- **Atmosphere/glass**: The bath atmosphere loaded with sulphur and SnS is under pressure, and finds its way out from the exit. This atmosphere is at high temperature and flushes the top and bottom surface of glass imparting a cloudy appearance.

The role of LORs on the quality of glass, especially the bottom surface, cannot be underestimated. The surface quality of the LORs should always be at a perfect state, no flaws of any kind can be tolerated. It is worth to mention again that as the glass is in the viscoelastic state, harm to the surface will be unavoidable.

In the first generation of float lines, including the LORs there were 226 rollers in the annealing lehr. Their Cr/Ni contents were different but within tight specifications. The diameter of the rollers decreases (305, 216, 203, 150 mm) and are different along the annealing lehr [63]. The rollers are found in two different zones in the lehr with different surface quality, Table 24.

| Zone | A | | B | C, D, E | F |
|---|---|---|---|---|---|
| Diameter mm | 305 | 305 | 216 | 216 | 150 |
| Metal | 310 S 25/20 | 302-18/8 | 302-18/8 | 25/4 | C 32 |
| Surface quality | 0·2–0·3 μm | | 0·8 μm | 1·6 μm | 3·2 μm |

Table 24: The diameters, the chemical composition (Cr/Ni) and the surface quality of rollers.

If the above indicated surface quality of rollers is maintained, then the bottom surface quality of glass will be 4 Å and the top surface will be 8·5 Å.

The glass follows a 40° angle as it is transferred onto the LORs. This angle reaches 0° at the 13th roller. The pull of glass is realised with the reverse rotation of the first roller, a point which is highly important. In general glass goes over to the third roller without touching the second roller. The roller temperatures must not change from bottom to top, otherwise eccentric rotation may occur, which negatively affects the quality of glass. Velocities in excess of 26 m/s may cause some vibrations in the rollers, leading to swaying and distortion of glass ribbon.

### 6.9.1 Bottom Surface Defects

Bottom surface defects arise because of the contact of glass and LOR, which subsequently weakens the surface properties. To prevent or minimise this negative development on the bottom surface $SO_2$ is introduced beneath the glass as it comes out from the exit.

$SO_2$ reacts with the surface of the glass at 580–600°C where $Na_2O$ of glass reacts to give $Na_2SO_4$, thus a cloudy appearance is formed.

The $Na_2SO_4$ is in the form of a powder, which acts as a kind of lubricant. By this mechanism the mechanical resistance of glass is not allowed to deteriorate.

If the conditions are not suitable several scratches may form on the surface. Different terms (impact check, chips, digs, cracks, scratches, etc) are used to describe the extent of the scratches. Any mechanical impact with glass can cause problems. The sulphate powder transferred to the lehr forms what is called 'sulphate rings' on the rollers. Several defects like glass particles, cassiterite crystals and metallic tin (related to the dross) may easily be embedded in these rings.

### 6.9.2 Sulphate Rings on LORs

Free particles (glass, etc) which are present in the environment may stick to the surface of rollers and cause serious harm like chips and scratches on the surface of glass. These types of defects can be observed under edge light conditions.

During cold repair these sulphate rings have been investigated in detail using micro-analytical methods. In the first three rollers more than 90% of the material is made of $Na_2SO_4$ the rest is gypsum ($CaSO_4$), cassiterite and glass particles Figure 71.

| $Na_2SO_4$ | Cassiterite ($SnO_2$) | Gypsum ($CaSO_4$) | Glass |
|---|---|---|---|
| >90% | 2–5% | 2–3% | 1–2% |

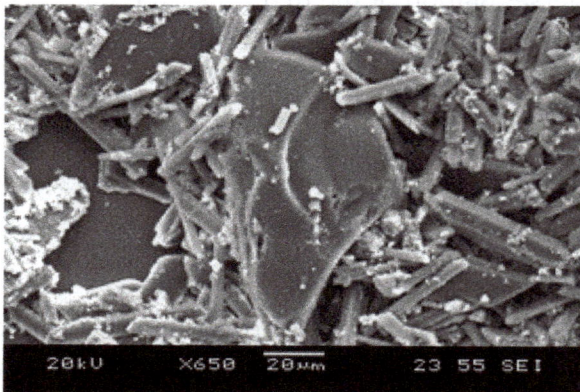

Figure 71: The particles embedded in sulphate rings. Long crystals are gypsum, the white particles are cassiterite and the angular grains are glass.

$Na_2SO_4$ powder acts as a binder for other particles. Cassiterite and glass particles have similar Moh's hardness to glass, therefore they can scratch the bottom surface. Cassiterite decreases in amount along the lehr, whereas Na-sulphate increases in proportion.

Figure 72: Sulphate rings on rollers just after the LOR.

This condition forms a risk for glass quality. Sulphate rings are quite hard in the hot region, where the thickness of the rings is about 5–8 mm, and tend to decrease along the lehr, Figure 72. An interesting feature of these rings is that most of them are at 3–5° angle to the axes of the roller, and oriented towards the left side. This off-centre situation of the rings may exert some strain to the glass, even causing some distortions in thin glass. The size of the sulphate rings in the hot part increases the diameter of the roller by 2%, which in turn may cause a 2% difference in velocity of glass.

# CHAPTER 7

## 7.1 Negative Developments in Glass Products after Production (Postproduction)

As noted several times before, in the many stages of glass production starting from the batch house to the warehouse, problems of varying degree are likely to happen. During this time defects may appear in glass products and losses of varying magnitude may be incurred. In other words, problems may arise in glass during and after production. It is quite frequent to see that some characteristic for the warehouse conditions. In the stock area glass comes into contact with several materials like paper, cardboard, nylon, plastic, rubber, polystyrene, textile felts all of which may cause some problems. For example, paper (newspaper quality) used as an interleaving material should have suitable specifications, such as 5–6 pH, moisture absorbance, etc. It is not infrequent to see hard particles such as quartz and feldspar in paper entrapped during production. Resin from paper is an obstinate material which sticks to the glass, not easy to remove.

Another point worth a mention is the cardboard used in several forms (box or plane) may have a reasonable amount of sulphur in its composition, this may cause surface deterioration. If these boxes are to be used they should not be wrapped with impermeable nylon. The boxes should be aerated with some holes. It is absolutely essential to know what the material may contain that will be in contact with glass. Filler materials used in paper and cardboard production may bear a risk.

Some exquisite glass products, like crystals, are very valuable for the owner and they usually tend to be overprotected by impermeable nylon. This is not very healthy for the glass product; corrosion may quite easily occur. The glass producer does their best, right from the beginning, to design a glass composition which will endure the negative environmental factors. Therefore, the customer should be informed about the best way to use this product.

## 7.2 Warehouse Conditions and the importance of Stock Parameters

All glass products are likely to spend some time in the warehouses of the glass producer and the customer. The longer the glass waits in the warehouse the more sensitive it becomes to corrosion. The main objective should be to deliver the glass products to the customer in the best condition possible, all quality criteria should be satisfied at the end of this transaction. The following analogy describes what is to be expected: Having produced the eggs in the farm, then the main objective should be to deliver the eggs to the customer in the best state possible (all fresh, all same size, no broken or fractured eggs, etc). They should not be overdue at all. It is the responsibility of the customer to keep the goods in the best condition.

The physical conditions in the warehouse should be appropriate; temperature and humidity variations should be within tolerable limits. Well aerated stock area is a key factor, preventing the condensation of vapour in the atmosphere.

The temperature at which condensation takes place is known as the **dew point**. For best results there should not be humidity condensation at warm and cold cycles. The humidity of atmosphere in the warehouse should not exceed 80%, in high humidity environment glass products are very sensitive for corrosion. Another important point is that the warehouse atmosphere should not be stagnant. An acceptable level of turbulence in the atmosphere is preferable.

To grasp the importance of parameters like temperature and humidity with time a satisfactory system for recording should be set up. This will provide valuable information for the variations for day, week, month and year. At best, the information about the fluctuations of temperature and humidity during a 24 h cycle are important for those who work in the warehouse and those who are in the transport of glass products. The comments made above are valid for all warehouses.

In a bottle container factory the bottles are stacked on one another on a pallet to form a multilayer with cardboard in between. This array of bottles is then wrapped up with an impermeable polymer. When a pallet is investigated, it can be seen that the bottles on the top and bottom layers may behave differently.

Glass bottles can corrode, to different degrees, more often higher towards the top of the pallet. The bottles contain a volume of air, which has some moisture in it. If the dew point is exceeded there may be a condensation. It is therefore logical to provide aerated conditions to the pallet to avoid condensation.

Glass with weak durability is likely to undergo corrosion. If the residence time of the stock is high, the risk of corrosion is high. If there is notable condensation there may even be some micro-organism activity. Surely, this type of development is not one that will please the customer. The exhaust of the lift trucks may leave carbon soot stains in the environment.

## 7.3 Container Glass Production Defects

Container products are used in all aspects of our lives. Much of our food is preserved in these products; therefore they deserve to be of high quality and be improved continuously. Medical material (serum, medicine) are packed in various container types.

There should not be any elemental migration to its contents for which special measures may need to be applied. It is for this very reason that the chemical durability of glass is very important. The total of heavy elements (Cr, Pb, Hg and Cd) in glass composition should not exceed 200 ppm.

Quality investigations for containers should start from raw materials. Random controls are useful to maintain the quality. There may be several problems related to the quality of cullet. Traditionally there may be contaminants in cullet, the most frequent being metals and glass ceramics.

Glass containers used for medical purposes and food preservation must be of special quality. There are four types of glasses for various purposes. Alkali migration is highest in high type glasses:
- **Type I**   Neutral composition (borosilicate glass)
- **Type II**   Neutral surface
- **Type III**   Normal soda–lime–silica glass
- **Type IV**   High alkali content soda–lime–silica glass

After improving warehouse conditions, dealkalisation can be used to prevent corrosion and scaling of glass. The **powder test** is classical for determining the chemical durability of glass.

Freshly filled bottles have a pH=7·03–7·25, after 4–6 months corrosion may develop where pH rises to 9·20–9·50. During blowing of bottles the air must be dry and free of oil. The $Na_2O$ and $MgO$ must be as low as possible, while $CaO$ may be increased, in addition if there is some $K_2O$ in the composition, altogether may improve chemical durability. The polymer used in shrink wrapping should be aerated to prevent condensation. In open air storing 2 months may be maximum.

If there is a limited amount of corrosion the following precautions may be useful:
- Rinse bottle with 8% NaOH solution heated to 80°C, then wash and rinse,
- Rinse bottle with 3% HCl solution heated to 50–60°C, then wash and rinse.

In a newly started furnace it is common to start with some defects which decrease in time. Seeds and stones are not infrequent which may be due to insufficient temperature. At high temperatures refractories may release some bubbles. If pull is not stable and refining unsatisfactory then some persistent bubbles may occur. Satisfactory furnace observation is necessary, one can determine weak points in the furnace which affect melting.

There are several points which can be optimised in order to solve problems: adjusting temperatures, cooling efficiency, suitable alkali level in the composition, electric boosting if available, suitable granulation of raw materials and optimum pull level.

Defects found in the molten glass have similar behaviour and origin in float, container and tableware furnaces, except scale factor. Quality requirements may vary in these productions.

Even though the general glass defect level may be tolerable, this does not mean that there are no defects at all, small defects are there and present. If defects are collected according to a schedule such that they represent the production areas, these can be systematically investigated. In this way it is possible to establish a 'certificate of production' for any particular furnace. It is thought to be informative how quality has developed over the last two decades.

Manual sample collection, follow-ups in later shifts according to our established principles were all carefully carried out. The defect intensity (%) were calculated as follows. Once the proportion of stones is calculated, then other defects, e.g. bubbles can be calculated by the same approach.

$$\text{Stones } [\%] = \frac{100\left(n_\mathrm{m} + n_\mathrm{a} + n_\mathrm{n}\right)}{3ts}$$

$$\text{Stones/ton} = \frac{1440\left(n_\mathrm{m} + n_\mathrm{a} + n_\mathrm{n}\right)}{3tp}$$

Where; $n_\mathrm{m}$=number of stones in morning shift, $n_\mathrm{a}$=number of stones in second shift, $n_\mathrm{n}$=number of stones in night shift, $t$=sampling time in shift (min), $s$=number of gobs per min, and $p$=amount of pull (t/d)

**Cat scratches** can be present in container products to a varying extent. Although they do not affect the function of the product, they are aesthetically disturbing. More detailed information about cat scratches was given in the section on vitreous defects.

The forming activities of containers are an important stage in production. Forming starts with the gob getting into the mould, followed by blowing and transferring to the conveyor. It is at this stage glass comes into contact with metal, that is mould and gob path. There may be several surface defects from this contact. If the defects appear in the same place on the products in a repetitive manner than it is possible to localise the defect origin. The gob shape and weight should not change out of limit values. There is also a tolerance for thickness variation which should not exceed max/min<2·5/1.

The bottles should stay vertically upright with no inclination, i.e. gob centring should be near to perfect. This may affect any diameter variation, hence roundness. The volume of bottles must be stable. If the bottles are returnable the internal pressure limit is 21 kg/m$^2$. As almost everything on the filling line is done automatically there is no allowance for limit values to change.

The mechanical resistance of the products decreases steadily on the line. Contact and friction of products is possible at all glass production stages and also at the filling stages at the customer's site. All these reduce the mechanical strength. It is possible to minimise the effect of contact and friction at the mould design stage.

To minimise friction between the bottles SnCl₄ solution is sprayed on the bottles just before leaving the annealing lehr. The coating that results acts as a lubricant for the bottles when they touch. A point that is neglected frequently is that no precautions are taken for the vapours emitted. These vapours come into contact with air in the fume hood where the formations of cassiterite ($SnO_2$) is possible. This forms a layer on the lehr structure which may drop on the products. It is possible to see cassiterite stones in products. If the cassiterite layer finds its way to the cullet then there will be cassiterite stones when the cullet is used. Cassiterite as a defect in container production happens only in this way.

## 7.4 Free Particle Problem in Medical and Food Containers

The chemical durability of glass bottles can be improved by using $SO_2$ to dealkalise them; the glasses are Type II as a result of this application. The glasses are suitable for use in isotonic liquids and serums. These products are assumed to be strategic because they are related to health. Before filling, the bottles are carefully washed and controlled thoroughly with dedicated instruments, and also manual controls. The inside of the bottles may have some **free particles** frequently seen after filling. This problem is always on the agenda between glass producer and isotonic solution producer.

Raw materials used in preparing medicine and the water used are examined in detail by using optical methods. The solutions to be investigated are filtered by using Millipore filters. Particles retained on filters are examined under a microscope or electron microprobe. Size, shape, quantity, organic or inorganic species should be noted.

It is recommended to use a troubleshooting approach for identifying and solving the problem. It is logical for customer to be suspicious of bottles. Raw materials used in the production of glass are also questionable.

It is vitally important therefore to identify the particle which is causing the problem. Experience has shown that the hardness of water used for washing the bottles may be the culprit. The $Ca^{2+}$ and $Mg^{2+}$ ions present in water combine to produce $CaCO_3.MgCO_3$ (dolomite).

The size of crystals is usually between 100–150 µm. To trace these crystals to their origin the water line and the fittings must be carefully investigated. Almost identical crystals can be found in the tea pots in our homes.

This type of idiomorphic crystals (perfect crystalline shape) cannot be found in the ground dolomite raw materials. For further verification trace element analyses can be made if necessary.

Free particles can also be present in water, beer and beverage bottles and fruit juice containers. In these bottles dealkalisation is not applied.

The filling lines are of high capacity, therefore special methods should be developed for quality control. The particle size is scaled between **S0** and **S5** as shown in the following Table 25. Most of the customer complains (80%) fall within S0 and S2 range [89].

| | Free particle classification | | | | | |
|---|---|---|---|---|---|---|
| Class | S0 | S1 | S2 | S3 | S4 | S5 |
| Size (mm$^2$) | >25 | 15–25 | 6–15 | 2·5–6 | 1–2·5 | <1 |

Table 25: The classification of **free particles** encountered in medical containers.

## 7.5 To Increase the Chemical Durability of Container Glass

The majority of container glasses are of soda–lime–silica composition. The bottles used for medical purposes are subjected to $SO_2$ gas application to improve the chemical durability (dealkalization) before they enter the lehr [64], Table 26.

The bottles which come into contact with $SO_2$ at about 600°C adopt a cloudy appearance, much like bloom. Investigation of cloudy surfaces shows the presence of minute crystals of $Na_2SO_4$, which are a couple of microns in size. $Na_2O$ is extracted from glass:

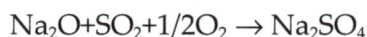

$$Na_2O + SO_2 + 1/2O_2 \rightarrow Na_2SO_4$$

Containers to be used for blood or fluid medication must be washed with water before final filling. Rinsing the bottle with water removes the $Na_2SO_4$ formed on the surface, which essentially dealkalises the surface and causes an enrichment of $SiO_2$, Table 26.

Under normal production conditions the containers produced are in **Hydrolytic Type III.**

| Glass containers > soda–lime–silica | |
|---|---|
| **Application** | **Treatments** |
| • Inside surface of beverage containers<br>• Inside surface of pharma vials and ampoules | • "Sulphur"; $SO_2$; sulphur dioxide<br>• "Fluorine"; fluorocarbon<br>• Ammonium bifluoride/ sulphuric acid<br>**>> Dealkalisation** |
| • Outside surface of beverage and other containers intended for high speed filling lines | • Hot-end coating<br>• Cold-end coating<br>**>> Strength retention** |

Table 26: Glass containers used for beverages and medical purposes are treated with acidic $SO_2$ gas to form $Na_2SO_4$, which is subsequently washed and rinsed; thus creating $SiO_2$ enriched surface (**Type II**). $SnClO_4$ is applied to form a lubricant external surface.

After **dealkalisation**, as mentioned before, the products are up graded to **Hydrolytic Type II**. This improves the corrosion resistance of products. Some improvements can be made by adjusting the chemical composition. For instance, if volcanic glass (perlite) is used instead of feldspar as a source of $Al_2O_3$ the persistent chemical durability problem encountered for amber glass will be remarkably improved. It is believed that the presence of $Al_2O_3$ and $K_2O$ together lead to this positive contribution, Figure 44.

The $SO_2$ gas in the bottles disperses into the lehr atmosphere, which causes some deterioration in the metal structure in the form of rust. If any turbulance happens in the atmosphere, then it is likely some of the rust particles will fall into the bottles, which may cause rinsing difficulties because they may adhere to the surface. The bottom of the bottles, as they travel on the conveyor chain may be smeared with this rust.

If the conditions in the stock area are not suitable, corrosion of glass products is possible. Complex Ca,Na sulpho/carbonate crystals are formed on the inner surface, Figure 73. In bottles used for alcoholic drinks, the pH of the liquid increases (>9) with time, in this case Ca,Mg-silicate flakes can form on the inner surface of bottles [65]. As pointed out before, if $Al_2O_3$ and $K_2O$ are used together the durability problems will be overcome [66,67].

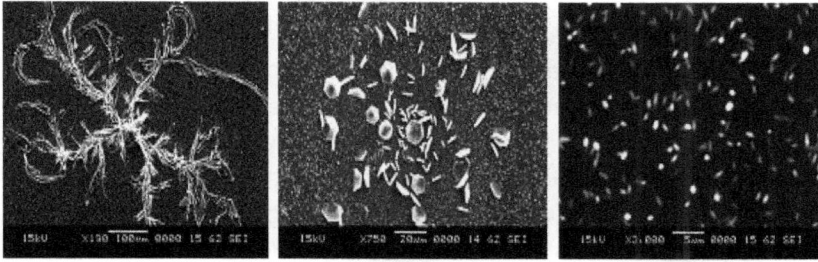

Figure 73: Corrosion is a frequent phenomenon encountered in container glass items. Complex Ca,Na sulpho/carbonate crystals are formed in the inner surface of the containers, the size varies from 2–3 μm to about 10 μm.

Because of high alkali content ($\sum R_2O$) in tableware products corrosion is more likely to occur compared to traditional soda–lime–silica products. In these products (automatic and handmade) dealkalisation is not applied.

Tableware products, due to long period of waiting can be subjected to corrosion in the stock areas of customers. The most frequent complaint for tableware products is their **behaviour in the dish washing machine** [68,69]. Although detergents may have some contribution, the main problem seems to be related to the hardness of the water. If the water is very soft the problem is more pronounced. The $Ca^{2+}$ and $Mg^{2+}$ ions in the water tend to prevent the replacement of ions in the glass. Thus the low hardness of water is a serious cause of corrosion in dish washers. Some inhomogeneities in glass become visible with the number of washing cycles.

**Decorative paints** on glass products are very susceptible to corrosion. Here, the colour of the print deteriorates first. The firing quality of the coloured **enamel prints** is very critical. If the firing is insufficient the problem shows itself in the early stages. Each enamel paint has a critical firing temperature. It is inappropriate to think that one set value for the decorating furnace will be enough. There is no one single standard firing temperature for paints.

A test in a gradient furnace can give the temperature at which the paint is bright and glossy. The decorating furnace must use this temperature. The PbO and $SiO_2$ contents of the enamel paints and the content of the pigments play a critical role on firing quality. The vitrification temperature increases with the silica content, while increase of PbO decreases the firing temperature. The stability of the decorating furnace is also important.

Tableware products have a relatively low **Vickers hardness**, therefore they can be scratched during handling. Of special importance is the negative effect of washing sponges on glass. Quite frequently there may be hard minerals like quartz, feldspar and corundum in the green part of the sponge (the Mohs hardness of quartz is 7, that of corundum is 9; glass has a hardness of 6·5–7). The two materials should be compatible in daily use. But the reality is that the green layer of washing sponge harms the glass surface [70]. It is the author's proposal that the glass producers and sponge producers should cooperate for mutual benefit. Again it is the author's opinion that, there is no need to use any material that is harder than glass for cleaning purposes.

## 7.6 Postproduction Problems in Float Glass

Most flat glass is produced by float glass technology. Having completed the forming operations in the tin bath the glass is transferred on to the lift out rollers (LOR) then proceeds through the annealing lehr. Float glass producers start to take several measures in the lehr up to the cold end in order to produce the best quality glass. The most popular method used is the application of acidic gas $SO_2$ on the top surface at a temperature 550–600°C. The $SO_2/SO_3$ reacts with the $Na_2O$ of the glass to form $Na_2SO_4$.

The purpose of this reaction is to remove $Na_2SO_4$ from the surface and leave behind a $SiO_2$ enriched surface. This is a typical dealkalisation reaction. The surface gains a cloudy appearance, like bloom. The glass treated with $SO_2$ has a higher corrosion resistance; therefore it can be safer in the stock area with atmospheric humidity.

The $SO_2$ applied under the bottom surface as the glass comes out of the exit has a different function than that blown on the top surface. The amount of $SO_2$ used for the top surface should not exceed $1·5$ m$^3$/h.

The water resistance of the glass surface is described as the amount of $Na_2O$ extracted from 1 dm$^2$ glass surface. If the glass surface is dealkalised then 80–100 mg/dm$^2$ of $Na_2O$ is a good value. The lower this value, the higher the water resistance of glass surface. Lower values should be preferred. The water resistance of the bottom surface is almost two times better than the top surface, which is undoubtedly due to the presence of $SnO_2$ in the bottom surface. Colourless float glass has a better water resistance than coloured glass.

Tempered glass can be doubly affected by water compared to normal glass. In other words under the same environmental conditions tempered glass is likely to undergo corrosion twice as much when compared to plain glass.

Float glass subjected to acidic gas ($SO_2$) application must be washed before cutting. Washing the glass surface improves the quality of glass. Glasses that are to be used for coating, automotive and mirror have to be washed and cleaned. Glass processors including tempering prefer glass to be washed on-line. This means that more or less all glass must be washed. In this sense, the washing system and the water quality must also be important parameters. Glasses that waited for a long time may need to be washed before transporting to the customer.

Washing the ribbon becomes necessary for satisfactory defect identification by the scanner system. Both surfaces need to be washed for the best results. The water quality used for prewash and rinsing is different. In this sense the water quality for float glass is important.

The **hardness of water** is a critical parameter. Water hardness is the sum of $Ca^{2+}$ and $Mg^{2+}$ ions. Silicate salts are more harmful for glass surfaces. Silica salts in excess of 2 ppm are difficult to clean. Therefore, there should be no droplets on the glass. Water used for prewash can be tap water, but that used for rinsing must be very soft and its silica content should not exceed 100 ppb. The quality of rinsing water is as follows, Table 27.

| $^0$d/H (Hardness of water) | | Water quality for mirror and coating glass | |
|---|---|---|---|
| Very soft | 0–4 | Conductivity ($\mu$S/cm) | <1 |
| Soft | 4–8 | Hardness ($CaCO_3$, ppm) | 0 |
| Medium soft | 8–12 | Silicic acid (ppb) | <100 |
| Hard | 12–18 | Organics (ppb) | <1000 |
| Very hard | 18–30 | Microorganism | <1000 pcs/l |
| | | Suspended particles | |
| | | <0·5 $\mu$m | <1000 pcs/l |
| | | <0·2 $\mu$m | <1000 pcs/l |

Table 27: The surface cleanliness of mirror and coating glass is of paramount importance, therefore the quality of water (rinsing) has very tight limits. On the left side of the table a general classification for hardness of water is given.

Glass which has been washed on-line and stocked in the warehouse may need some further protection from corrosion. If the transport of the glass is to be made to places where humidity is high, then it is possible to apply a liquid acidic protector on the surface at 40–45°C, just after the glass cutting zone. It is absolutely necessary that this acidic layer must be dry before stacking. If it is not dry the suction cups may lift some obstinate imprints on the glass. Suction cup rubber material should carefully be investigated for its oil and filler type quality. Some traces on the surface of glass may be due to suction cups. To prevent such traces, textile covered suction cups have also been used with considerable success. For best performance the atomised acidic liquid spray must be dry.

### 7.6.1 The Protection of Float Glass and Problems Arising in Storage

The sizing of float glass pieces is accomplished by automatic systems. The smallest sizes are cut manually. At the edges cut by X–Y systems very often there may be glass particles. The cutting oils used must be water soluble and able to evaporate quickly. This will prevent the glass particles sticking to the surface. The oil trace on the glass should not exceed 10 mm. The fresh glass particles strongly adhere to pristine surfaces. These particles may frequently be loosened during transport and cause severe scratches, which may then go on to form weak points for corrosion.

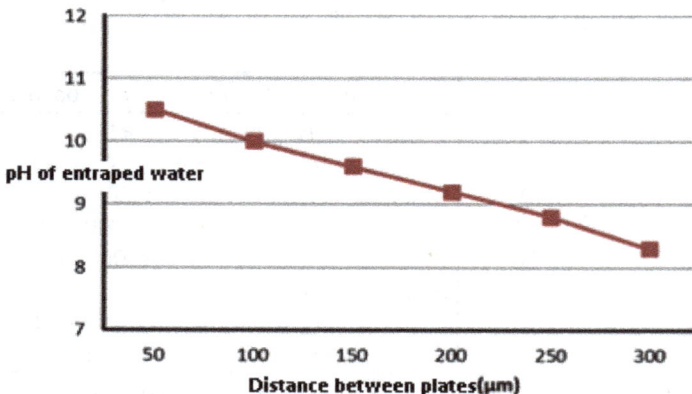

Figure 74: The distance between the glass plates has a direct influence on the pH of condensed water between the plates. If distance is less than 200 μm, then pH>9 can easily be realised, which is critical for corrosion.

After glass is automatically cut an organic powder called leucite is sprayed on the top surface. The average size of this powder is about 100 μm, but different sizes may also be used. Too coarse particle may create glass breakage! The size of the powder should be such that no condensation of humidity is possible. Otherwise if there is condensation pH value may increase to 9, which is critical for corrosion, Figure 74.

Additional protection for glass in the stock is necessary. Adipic acid is known to hinder corrosion, therefore it can be used as a mixture with the leucite powder in proportions of 10, 15 or 20%. To use of adipic acid can be cost effective. This acid has a low solubility in water. Leucite acts as an interleaving and protecting material against scratches. The powder is rendered electrostatic so that it sticks to the surface of the glass. If this is not done satisfactorily the powders will fall off the surface during transport. In extreme cases the glass surfaces may stick together if there is no powder left. For large size glass plates waiting in the warehouse moisture absorbing paper (newspaper quality) or leucite must be used. Groups of glass plates (10, 20, etc) may be separated by using polystyrene plates or felts of suitable size.

Glasses which wait a for long time in storage must be controlled carefully before transport to the customer. If this is omitted some corroded glass may, by mistake, be sent out. This is a high cost mistake for the glass producer, which may lead to losing the customer. All types of packaging used for transporting the glass must be mechanically strong, therefore a good deal of engineering is necessary.

The wood used for constructing the end-caps, etc, must not lead to any micro-organism activity. To be on the safe side, the wooden structure is subjected to fumigation (application of acid fumes) before export.

Corrosion may take place over short or long times in the glasses in the stock area. The cause behind this is the moisture condensed between the plates as the pH may have exceeded the value of 9 and could easily trigger off corrosion. This is accelerated if there are high and low temperature cycles in the environment. It is important that the relative humidity of the warehouse should not exceed 80%. It may be required to put some additional precautions in practice to suppress the humidity. The top surface of float glass is more sensitive to corrosion. This type of corrosion can be observed easily on the pallets. The alkali and alkali earth ions tend to move out from the corroded parts of glass. In the final stage the glass becomes weakly semi-transparent. As these elements move out from the glass they form hydroxides of Ca and Na. In addition to hydroxides there may also be some carbonate species because of the presence of $CO_2$ in the atmosphere. It is quite likely that in this environment the pH may also increase due to cyclic conditions.

Some glass producers go one stage further for corrosion protection. They may put some pockets of silica gel, clay or talc in suitable places on the pallet. If they are open to the atmosphere there will be no advantage at all. During handling these pockets may burst on the glass, which is impossible to clean.

### 7.6.2 Particles Accumulated on Float Glass Surfaces

As the glass ribbon travels in the lehr, there is always a possibility that some particles may fall on the surface, or the glass may come into contact with gases. Foreign particles may fall on the surface as the ribbon is still on the line. The following points are noteworthy:
- Gases and vapours in the environment: $O_2$, $N_2$, $CO_2$, $SO_2$, HCl and $H_2O$ vapours. All these may be physically adsorbed on the surface,
- Particles present in the atmosphere. The application of $SO_2$ at the entrance of lehr may give rise to some metal–gas reactions whereby rust particles fall on the surface,
- Crystals of $Na_2SO_4$ are always present if there is dealkalisation,
- Traces of imprints left on the surface of glass by interleaving material (leucite, paper, etc),
- Traces of cutting oil on the glass,
- Several handling signs like finger, label, suction caps, roller imprints can always be found on the glass.

Contaminants of any nature on the glass surface must be able to be cleaned without leaving any trace on the glass, otherwise a coating of any kind will face several problems. For glass to be coated the following precautions must be applied [71]:
- To use special gloves during handling is simple and effective,
- Suction cups must be covered with special textile to minimise the possibility of a chemical reaction between rubber and glass,
- Oils used must be water soluble and easily evaporated,

- As the glass enters the coating section there should not be any vibration of the metal system, all the surface must be homogenously clean.

## 7.7 Suction Caps and Glass Contact Effects

After glass is cut to suitable sizes, each plate is picked up by a robot and put on a stacker. The robots that lift the glass plates use a number of suction caps, the size of which may change depending on the dimensions of glass plates. Big suction caps are used for large size glass plates.

The suction cups are made of rubber of special quality. The elasticity, hardness, resistance to atmospheric conditions and type of filler used are some of the properties important for glass. The most important requirement is that the suction caps should not leave any imprint on glass surface and not give rise to any pressure failure.

The suction cups exert a considerable pressure on the glass; as the rubber behaves in elastic manner, there is always a possibility of some oil marks may be left on the surface. As the cups exert a pressure, it is possible that the leucite and adipic acid particles between rubber and glass will be deformed and stick to the surface. The imprint left by the rubber may not be easily cleaned by standard washing, they may be visible around dew point temperatures. It would not be possible to use this glass for any coating purposes.

The corrosion which develops in the region of suction imprints may behave differently; either more or less severe with no plausible explanation. Suction cup problems are frequently the topic of customer complaints. When investigated by microscopic methods, suction cup marks show the presence of a high concentration of almost dried oil and rust particles. Therefore, periodic cleaning may be useful, or even develop a special method to prevent this development.

It will be more appropriate for the glass producer to dictate their requirements to the suction cup producer. Always getting full information about the chemical and physical properties.
- Natural rubber must be used,
- Parafin oil content must not be less than 5%,
- Filler particles harder than glass (quartz, titanium dioxide, etc) must not be present
- Suction cups must be conditioned before use (7 days at 100°C).

## 7.7.1 Corrosion of Float Glass and Precautions to be Applied

Although glass is chemically highly resistant, it is not unusual to see that the glass may react with water, water vapour and $CO_2$, all of which can be found in the environment [72].

The most persistent contact is that of glass and the atmosphere. There may be changes in the surface properties due to this contact [73].

The effect of moisture is as follows: As glass comes into contact with atmospheric water vapour the alkali ions ($Na^+$, $K^+$) and the alkali earth ions ($Ca^{2+}$, $Mg^{2+}$) start to replace the $H^+$ of water or hydronium ion ($H_3O^+$).

These reactions are called **First Stage Corrosion Reactions.** The silica network of the glass does not change at this stage. But as the replacement reaction proceeds, that is as the $H^+$ ions are consumed by glass, the $OH^-$ ion concentration increases.

This causes the pH increase of the liquid. This is called the **Second Stage Corrosion Reaction**, where pH>9. At this stage the silica structure is disrupted and this is followed by the **Hydrolysis Reaction**, Figure 75:

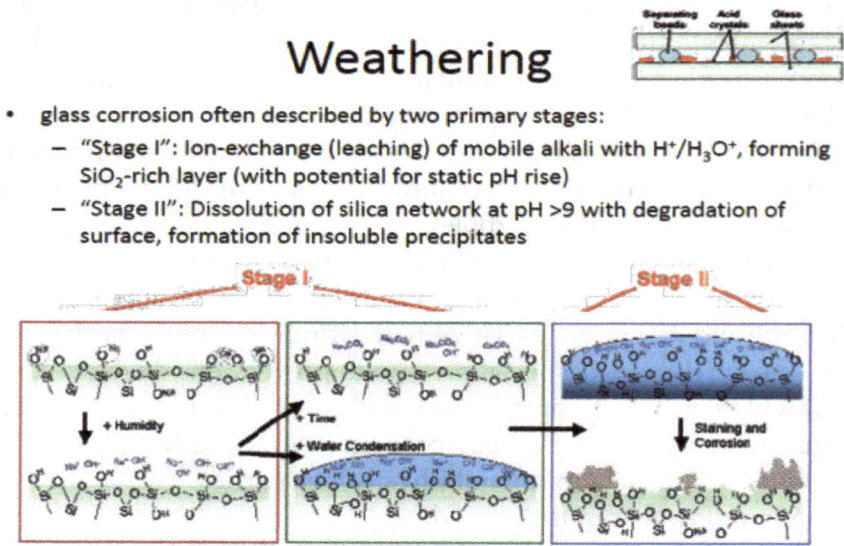

# Weathering

- glass corrosion often described by two primary stages:
  - "Stage I": Ion-exchange (leaching) of mobile alkali with $H^+/H_3O^+$, forming $SiO_2$-rich layer (with potential for static pH rise)
  - "Stage II": Dissolution of silica network at pH >9 with degradation of surface, formation of insoluble precipitates

R. A. Schaut, Pantano C.G., The Glass Researcher (2005)

Figure 75: Corrosion mechanism in float glass, the presence of humidity is essential, which condenses on glass under suitable temperature conditions. As pH>9, then Stage II reaction takes place which disrupts the surface of quality.

Glass corrosion is in a way a leaching reaction, the effectiveness of which depends on temperature, pH, glass composition and the contact period in the environment.

**Leaching Reaction**: There is a replacement reaction between the elements of glass ($Na^+$, $K^+$, $Ca^{2+}$ and $Mg^{2+}$) and $H^+$ or hydronium ion ($H_3O^+$) from water which leads to the formation of S–OH groups. The structure of glass is altered as shown in the following reactions:

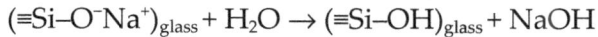

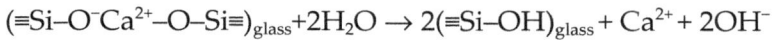

$$(\equiv Si–O^-Na^+)_{glass} + H_2O \rightarrow (\equiv Si–OH)_{glass} + NaOH$$

$$(\equiv Si–O^-Ca^{2+}–O–Si\equiv)_{glass} + 2H_2O \rightarrow 2(\equiv Si–OH)_{glass} + Ca^{2+} + 2OH^-$$

As a result of these reactions the structure of glass becomes weakened (leaching) in alkali and earth alkali ions. The leaching reaction slows down depending on the chemical composition of glass. At the end of this reaction the network formers of the glass keep the same position without any change.

## Effects of Glass Corrosion

- dissolution/etching/weight loss
- leaching/ion-exchange/surface layer formation
- hazing/ dimming/ pitting, staining/ VISUAL EFFECTS
- roughening/microporosity/ REACTIVITY
- increased susceptibility to soiling/difficulty cleaning
- STRENGTH and FATIGUE

Figure 76: Effects of corrosion on glass surfaces; dissolution and leaching by ion-exchange leads to visual effects like staining and pitting. The surface becomes rough and porous. At this stage cleaning of surface becomes impossible.

**Corrosion Reaction**: If the pH of the liquid exceeds 9 the leaching reactions on the glass surface slow down; at this stage $OH^-$ ions attack the Si–O bonds in the glass structure to cause a total solution.

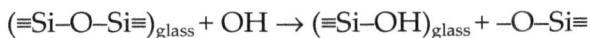

$$(\equiv Si–O–Si\equiv)_{glass} + OH \rightarrow (\equiv Si–OH)_{glass} + –O–Si\equiv$$

The reactions mentioned above (replacement and hydrolysis) may take place easily on the surface of glass even in stock conditions leading to the corrosion of glass [74–76], Figure 76. Even though float glasses are resistant to external factors, if the environmental parameters (humidity, temperature, packaging, etc) are not suitable the glasses will undergo corrosion.

Corroded glasses cannot be restored to their initial quality. It is more realistic to think that glass is not of infinite life. Beyond 75% humidity conditions glass will progressively loose its chemical properties [77,78], Figure 77.

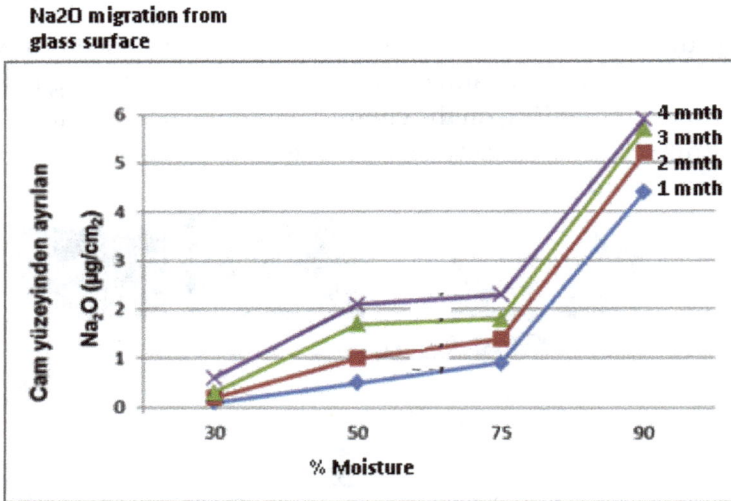

Figure 77: The threshold humidity value of 75% is critical. Above this value the pH increases (>9), as well as the acid consumption required for neutralisation.

Within the framework of what has been explained before, it appears that the main target is to have a strict control on the pH of the environment; never let it rise to a value of 9. There are some points which will positively contribute to this dilemma:

- The chemical durability of glass is closely related to its chemical composition. $1–1·5\%$ $Al_2O_3$ is a good value for durability in many parts of the world,
- Alkali neutralisation of glass surfaces should be realised by keeping pH on low side,
- Dealkalisation by applying $SO_2$ between 540–580°C results in

SiO$_2$ surface enrichment,
- Temperature and humidity control in the stock area
- Dust free atmosphere and emission free lift trucks in the warehouse,
- Use shorter stock periods,
- The distance between the plates should not be less than 200 µm, otherwise pH may arise,
- Improvements on stock and transport conditions should be made, for example using interleaving paper (pH=5–6), 10–15% adipic acid containing leucite, apply acidic liquid film and powder protection together [79].

If acidic liquid is applied to the surface, it must be borne in mind that the transmission of the glass will be reduced, this is a point where the customers are not very pleased.

In addition to surface protection studies, it is important to develop the chemical composition as much as possible, for example mixed alkali effect (Na$_2$O+K$_2$O) may be helpful and with a small increase in the batch cost (3%) chemical durability is improved by 10% [80].

It is the authors' firm belief that one day in the near future the glass products will pass through an electromagnetic field, so that the Na ions are strictly controlled in a way that will prove useful for glass corrosion [81,82].

### 7.7.2 The Effect of Water and Humidity on Glass Corrosion

Glasses that are installed in windows or building facades are in contact with water almost continuously during their use. As they are continually washed there is no corrosion, if there are no additional negative conditions. But glasses in the stock, if there is not enough spacing, because of humidity, can be at risk.

Water used for cleaning can cause some problems if the water is hard. This creates some obstinate marks as a result of evaporation. Water used in sprayers in gardens causes similar marks. If the hardness is very high (Ca and Mg are high) and if there is silica in excess of 10 ppm then the problem can be serious. If high silica containing water evaporates, there is no doubt, there will be marks left, which will damage the surface of the glass. It is for this reason that detergents containing silica should not be used for cleaning. These marks cannot be restored in any way.

In high humidity regions there is a high risk of condensation on the surface of the glass. If the temperature falls below the dew point, then all conditions are suitable for condensation. This problem is frequently observed during transport, in open atmosphere storage and in warehouses with poor air conditioning.

If the humidity is high around glass packages (stagnant atmosphere), if it comes into contact with cold glass condensation would be unavoidable. Glass temperature decreases at nights, and the whole mass of glass gets cooler. Because of the heat capacity of the glass, it will warm slower than the warm air around the glass. As the day gets warmer then evaporation will follow. In order to avoid condensation it is advisable to keep the temperature around the glass pallets well above the dew point. This can be accomplished easily in warehouses where there is heating. The table below shows the behaviour of humidity and temperature during glass transport, Table 28.

| Min. night temp. of glass | Max. day temp. of glass | Temp. difference (day/night) | Relative humidity in the day* | Condensation risk |
|---|---|---|---|---|
| –18°C | -7<br>4 | $\Delta T=11°C$<br>$\Delta T=22°C$ | 38%<br>16% | None<br>possible |
| –4°C | 7<br>18 | $\Delta T=11°C$<br>$\Delta T=22°C$ | 44%<br>21% | None<br>possible |
| 10°C | 21<br>32 | $\Delta T=11°C$<br>$\Delta T=22°C$ | 50%<br>25% | None<br>possible |

*If humidity exceeds these values during the day condensation occurs

Table 28: Temperature variations in day and night during transport.

In general there are some safety limits for daily variations in temperature and humidity where glass is stocked. For example day/night temperature should be less than 5·5°C and the humidity value during the day should not exceed 70%. If the temperature of the warehouse is higher than the outside temperature then there is no risk of condensation. However, if there is a high fluctuation in humidity, the glass temperature should not be below the dew point.

### 7.7.3 Why Humidity is Critical?

In its simplest form humidity is the amount of water in a given volume of air ($M_{air}$). However the temperature of air dictates how much water the air will contain at most ($M_{max}$). Listening to TV or radio everyday, what we must understand about humidity is that it is actually **relative humidity (= % $M_{air}/M_{max}$).**

As a general rule warm air contains more humidity than cold air. This is why, on hot days in the summer, the humidity content is more and therefore people feel uncomfortable.

## Weathering (leaching and corrosion by HUMIDITY)

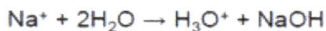

$$Na^+ + 2H_2O \rightarrow H_3O^+ + NaOH$$

These hydroxides then react with carbon dioxide from the atmosphere to form carbonates, as in the reaction

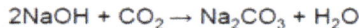

$$2NaOH + CO_2 \rightarrow Na_2CO_3 + H_2O$$

and the reaction

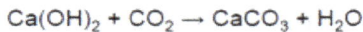

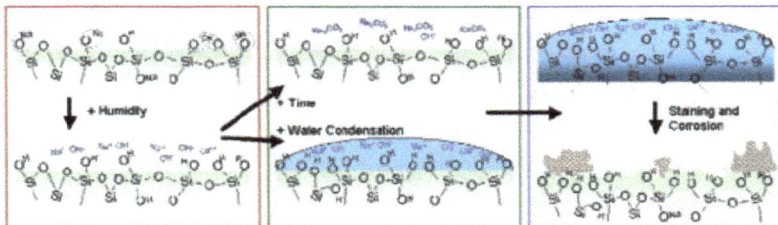

$$Ca(OH)_2 + CO_2 \rightarrow CaCO_3 + H_2O$$

Figure 78: Following the condensation of humidity hydroxides start to form, these then react with the $CO_2$ of air to form carbonates of Na and Ca as corrosion products.

From meteorological point of view **dew point** is just as important, the temperature where humidity is saturated and condensation commences, e.g. if the outside temperature is cold enough, then the humidity in our houses start to condense on the inner surface of windows.

This is very much like what happens when a water jug is taken out from the fridge, there is an immediate condensation (the humidity of environment) on the surface of the jug.

The conditioning of warehouses or depots becomes very important at high humidities. At high temperatures high humidity air condenses on the walls or floors of buildings. As mentioned before, if the humidity is more than 70% there is always a risk that all glass products could undergo corrosion, Figure 78.

Glass plants are usually located in barren places outside the cities. The meteorological activity (humidity, temperature, wind, rain, etc) on daily, weekly, monthly and yearly bases must be carefully evaluated. In fact this information is necessary before the plant is erected. There are some remarkable temperature and humidity variations in such a place [83], Figure 79.

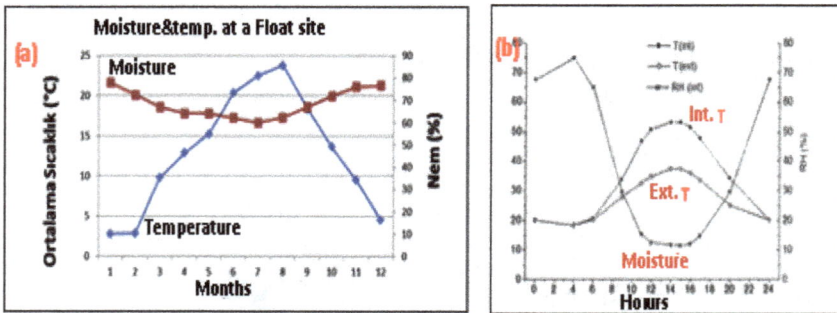

Figure 79: Annual temperature and humidity changes (a) and daily (24 h) changes (b).

The warehouses must be constructed in such a way that conditions for minimum condensation should be realised. The day and night temperatures must not fall below the dew point. The following table shows the amount of water content at 60% relative humidity and application temperatures of 25, 30 and 35°C. Values in parenthesis are for 80% relative humidity, Table 29.

|  | 25°C | 30°C | 35°C |
|---|---|---|---|
| Water content in air g $H_2O$/kg dry air | 0·014 (0·017) | 0·017 (0·023) | 0·023 (0·028) |
| Condensation temperature (°C) | 20 (22) | 22 (27) | 27 (30) |

Table 29: Water contents in air at three different temperatures compared with 80% relative humidity.

It is possible to calculate the amount of water content in the air entrapped between the glass plates of 1 m$^2$ area, a spacing of 100, 150 and 200 μm. Two values of relative humidity are taken 60 and 80% (the density of air is assumed to be 1·226 kg/m$^3$, and the size of jumbo glass 6×3·2 m), Table 30.

| | Distance between glass plates (μm) | | |
|---|---|---|---|
| | 100 | 150 | 200 |
| Water content (g/m$^2$) at 60% RH | 0·0017 | 0·0026 | 0·0034 |
| Jumbo (g/19·2 m$^2$) | 0·0329 | 0·0494 | 0·0658 |
| Water content (g/m$^2$) at 80% RH | 0·0021 | 0·0031 | 0·0042 |
| Jumbo (g/19·2 m$^2$) | 0·0399 | 0·0600 | 0·0799 |

Table 30: For two different relative humidities (60%, 80%) the amount of water condensed at three different spacings between glass plates. One can infer that there will be a higher concentration of NaOH in smaller spacings which means more severe corrosion.

The size of the condensed moisture between the glass plates is about 4–5 μm. Each of these points can be a starting point for corrosion, this is enhanced if the temperature of the warehouse is near or below 20°C. This point has to be taken into consideration for conditioning the warehouse. For those who are involved in stocking and transporting of glass. The following points are worth to keep in mind:
- The humidity of warehouses must be kept at acceptable levels; which can be achieved in several ways,
- To allow the temperature of the warehouse to increase in the summer aggravates the problem,
- Humid air atmosphere in the warehouse and the moisture accumulated on the walls of building should be expelled as much as possible,
- Fans should be used to stir up stagnant warehouse atmosphere,
- The insulation and air conditioning of warehouses should be designed accordingly,
- Evaporation from warehouse floors can be hindered.

## 7.8 Investigation Methods of Corroded Glass Products

### 7.8.1 Investigation of Float Glass

Surface quality loss in float glass may develop in several stages of handling. If there is any complaint by the customer all the details should be noted, for example production date, season, warehouse conditions, interleaving material, the intensity of defect and its distribution on the glass plate. For additional observations the glass should be cleaned by using compressed air. The first thing to be done is to identify the tin side. The interpretation of the defects should be linked to the surface at all stages. Irrespective of the size of sample the surface information must be noted.

There are several optical methods which can be used for investigation. First of all the glass plate should be examined under edge light conditions in order to locate the defect. The same observations should be made after washing the glass with distilled water.

The illumination power of the light in subsequent investigations is important. Corrosion and any surface defect are observed more effectively by this approach. For example, most of the time, the suction marks cannot be seen by the naked eye, but can easily be located by edge light investigation. Marks left by interleaving material (paper or powder) are also easily observed, which otherwise would be difficult. **Never underestimate the effectiveness of edge light investigation.**

Defects like corrosion, suction marks and paper marks can best be observed if the sample is subjected to vapour or steam. To strongly breathe on the sample is also effective for quick observations.

Float glass quality must be examined for '**ream**' by optical methods. The ream value is very critical for automotive glass uses. The sample is illuminated at an inclination in front of a white screen. Ream can be observed in parallel lines with varying strength.

For more detailed studies on ream, the sample can be etched chemically (HF, 15–30 s). The points identified as ream can be investigated using electron microprobe. After long counts it will be seen that ream has slightly higher $CaO$ and $SiO_2$ (about +0·5%). This shows that ream is a chemical inhomogeneity related to melting and not an optical distortion related to tin bath.

## 7.8.2 Investigation of Container and Tableware Products

Before cutting the sample it should be washed with clean water, followed by detailed defect investigations. Strong illumination should be used on dark background. During the observation the light should not come to the eye of the observer, the light should be positioned at the back of the observer. All details should be noted at this stage.

Before cutting the bottle or container it should be washed with **methylene blue liquid.** This die attaches to the micropores and renders the defect clearly visible. After cutting the sample all necessary observations can be made. The same methodology can be applied to tableware products.

The characteristics of defects can be explored by using sophisticated methods such as electron microprobe, XRD, atomic force microscopy, FTIR and XPS. In this way the chemical composition and crystallographic structure can be elucidated.

Although it was mentioned in some detail before, it is worth repeating that any analytical detail about cords, cat scratches and knots can also be obtained with these methods.

Although it is important to make detailed and comprehensive analyses, it is more important to put this information to practical use for possible precautions. If the warehouse conditions are satisfactory it is easier to improve the parameters. Warehouse design which takes into account the phenomenon of the corrosion of glass products at the initial stages of the investment will prove successful in the later stages.

**Special Note for Float Glass Protection:**
**Extensive use has been made of the Technical Material prepared by Pilkington for customers, especially for float glass subjected to corrosion under stock conditions. A summary of this will be found in the following pages.**

## 7.8.3 Protection of Glass Surfaces

To prevent the deterioration of glass surfaces the first thing to do is to not allow the drying of water or any chemicals on the surfaces. Equally important is that condensation on the glass must be prevented at all costs. The temperature of the warehouse and stock areas must be stable and well above the dew point and the humidity of air must be less than 80%. In glasses stocked outside there is always a risk of condensation. This is due to the fact that the bulk mass of glass is of lower temperature than the surrounding air. Cyclic temperature and humidity conditions will cause problems. Corrosion marks left behind on glass surfaces are very difficult to clear and very costly, not worth the glass itself. It is much better not to allow the glass to deteriorate, than try to restore it by means of polishing and chemicals.

### 7.8.4 Suitable Stock Techniques

The following points should be taken into consideration for glasses in warehouses, in open stock conditions and in transit:

- In warehouses and stock areas glasses in pallets and end caps may be subjected to corrosion if the spacing between the glass plates is not enough. There should be an interleaving material in any case,
- Effective air conditioning is a prerequisite for the prevention of corrosion. Fans can be utilised effectively. Enough separation between pallets and end caps is necessary. Warm air should not be allowed to enter the warehouses.

### 7.8.5 Cleaning of Corroded Glasses

If corrosion is severe it is impossible to restore the glass surface to its initial quality. But in the early stages of corrosion some trials may be attempted:

- Cerium oxide paste can be used for polishing,
- Use of $HF+NH_4F$ mixture with a stable pH can be tried on the glass; different application periods (5, 10, 15 s) can be tested.

Glasses should be cleaned with soft textile by liquids which are not basic chemically. The glass should be rinsed as quickly as possible with the aid of a blade. If there is oil and glass mastic marks on the glass then xylene and toluene can be used for cleaning followed by rinsing and drying.

Glasses can be scratched considerably by washing sponges (the active green layer may contain quartz and corundum). Red and green coloured washing sponges should not be used to clean glass.

## 7.9 Secondary Processing of Float Glass

In various stages of float glass processing, e.g. tempering or heat treatment may cause distortion. As glass is heated up to the softening temperature there may be some sagging as it travels over the rollers, Figure 80. These are known as **roller marks** and appears in the form of undulation distortions. Sagging is maximum at the free edges of the glass. Buildings in which these glasses are used may show a very disturbing appearance in reflective mode.

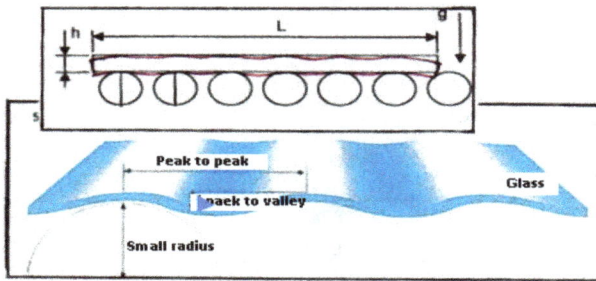

Figure 80: Float glass is likely to undergo some degree of sagging as it travels over the rollers. The edges show most warping.

If float glass is subjected to any kind of heat treatment there will always be some loss in quality. It is best to keep the optical distortion effect at minimum.

Ideally in buildings with heat treated glasses the images of reflected objects (buildings, clouds, trees and pylons) should not be distorted; but this is almost impossible. Similar views can be observed in all countries, some severe some tolerable. It is extremely unfortunate that float glass of top quality can end up in this form after processing. As the softening temperature of glass is within a narrow range, heating and cooling should be made as quickly as possible to overcome this problem. This is considered to be a weak point of glass processing.

Float glasses used in buildings can be processed in various ways [84]:

- Normal annealed glass,
- Tempered glass (toughened)
- Heat treated glass,
- Laminated glass.

Defects encountered in these products can occur in the warehouses and during use. The customer can get in contact with the processors and finally to the glass producers, which provides a chance to improve any inefficiencies at any of the stages. The properties of these products are summarised in Table 31, [85].

| Properties | Annealed glass | Heat treated glass | Tempered glass | Laminated glass |
|---|---|---|---|---|
| Wind load | Raw glass strength (1×) | For same thickness strength (2×) | For same thickness strength (4×) | If heat treated (1·5×–1·8×) If tempered (3·0×–3·6×) |
| Impact resistance | Medium | Higher than annealed glass | Higher than annealed glass High security glass | If tempered like 'security glass' if not medium |
| Breaking property | Large. long, Sharp edged, many pieces | Breakage less, pieces rather large | Glass breaks into small pieces | Radial breakage from impact point, due to PVB, glass stays intact |
| Applications | Architectural, Safer in thick glass, weak to hard impacts | Suitable for shadowing effect and for spandrel, more resistant to breakage, very low distortion | Distortions in reflective mode, very resistant to breakage, if compression layer is not scratched no breakage, spontaneous breakage due to NiS | Architectural, Automotive wagons, boats and ships. Furniture |

Table 31: Several properties of annealed, heat treated, tempered and laminated glasses are brought together.

### 7.9.1 Spontaneous Breaking of Tempered Glass

Very small size (40–80 μm) nickel sulphide (NiS) particles in tempered float glasses may create very serious problems after installation in buildings. There is no consensus on the source of NiS particles [84,86].

At about 380°C, NiS undergoes a phase change, where there is a considerable volume decrease which leads to spontaneous breakage of the glass.

To prevent this damage in tempered glasses all glass units are subjected to a temperature increase of up to 380°C with a suitable soaking time, to cause the breakage of those units which contain NiS.

This process causes some high costs, but for safety reasons it is accepted to be the only way out (90% success). Glasses installed at an angle bear more risk.

| Properties | Annealed glass | Heat treated glass | Tempered glass | Laminated glass |
|---|---|---|---|---|
| Preliminary preparations | -Furnace observation and imaging<br>-Tin surface<br>-Edge light examination<br>-Use different lighting conditions<br>-Apply optical methods,<br>-Use SEM or AFM<br>-To cut with diamond cutter is easy | -Determine the tension<br>-Observation and imaging<br>-Tin side<br>-Edge lighting<br>-Apply optical methods, use SEM or AFM<br>-Cut carefully with diamond cutter | -Observe with polaroid glass prior to tension release<br>-Heat to 400–500°C to release tension<br>-Possible to cut with cutter<br> -If the sample is too big, wrap with nylon,<br>-Mark point to observe, then examine | -Observe if there is vacuum<br>-Localise defect<br>-Heat sample to 240°C for separating PVB<br>-Start optical observation<br>-If any particle in interface localise and investigate<br>-Not possible to cut if tempered release tension |

Table 32: Investigation methods for annealed, heat treated, tempered and laminated glass.

**Before releasing the tension of tempered glass, observation with polaroid glass may be very useful to identify the high stress points.**
After this stage, the glass can safely be cut or broken. The breaking pattern of tempered glass should be carefully examined with a loop or stereomicroscope and compared to literature findings, Table 32. Although it is a tedious job, if examined with patience, it may be possible to find the culprit.

### 7.9.2 Quality Problems Related to Insulation Units

With energy rating highly in the priorities for building designs, insulation units find extensive use in windows. It is expected there will be greater use of insulation units as they provide a clear advantage on energy consumption and environmental grounds. This alone creates a big advantage for glass producers. The quality of float glass used in insulation units is protected by using BS 5713.

Glazing type distribution in the EU

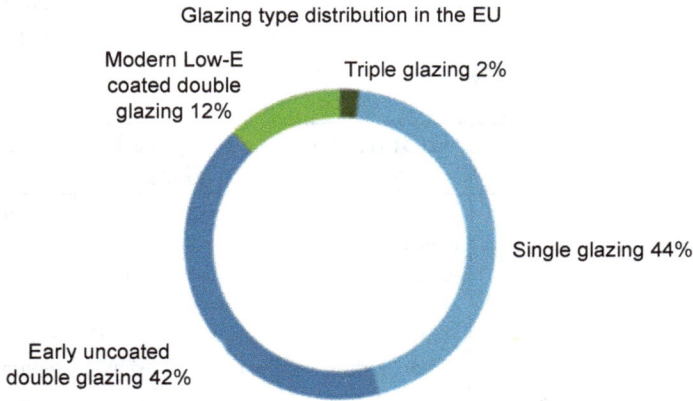

Modern Low-E coated double glazing 12%

Triple glazing 2%

Single glazing 44%

Early uncoated double glazing 42%

Figure 81: The use of insulation units in Europe.

Insulation units make up 50–60% of the windows in Europe, Figure 81. In the last couple of decades multiple glass panes made up the insulation units, with special filling gases extensively used, achieving much higher energy performance values.

Insulation units are made of at least two panes. After they are put in metal frame a special mastic is used to avoid any air ingress. Silica gel keeps the air dry in the unit. The quality of all the materials used in producing insulation units are controlled by different specifications to ensure long life (about 10 years or more). Sometimes poor workmanship may cause some problems such as air inlet. If air goes into the units then condensation can occur in the inner surfaces of glass, which eventually can give rise to corrosion.

If coated glass is used under unsuitable conditions, the moisture can react with the coating causing some decolourisation. This type of development may also occur in shadow.

Observations of insulation units with the unaided eye is essential before the unit is dismantled. Micro analytical methods must be applied at this stage. Various illumination styles may be useful.

Some customers may complain, though not very frequently, about a 'rainbow colouring effect' appearance in glazing units. It varies from observer to observer and is essentially an optical illusion. It is possible to find a solution to this, but not very practical for production. If glasses cut parallel to the pull direction are used in the same unit then this effect is visible, on the other hand if glasses cut perpendicular to the pull are used in the same unit, this rainbow effect is not visible. This is explained in Figure 82.

**GLASS RIBBON**

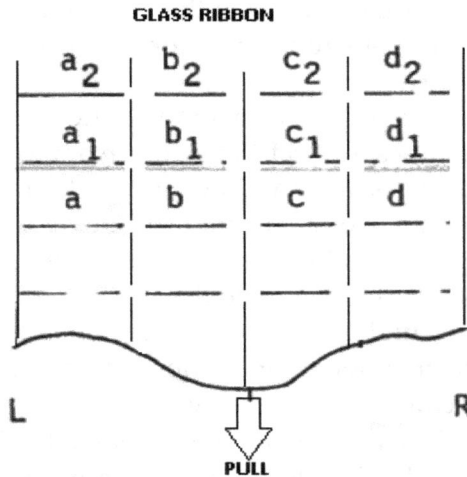

Figure 82: The 'rainbow effect' is frequent in double glazing, it is considered to be inherent. If the glass sheets following each other, parallel to the pull (b, b1, b2) are used then this effect is seen, but if glass sheets perpendicular to pull (a, b, c, d) are used the rainbow effect does not occur.

Insulation units produced with plastic profile (PVC, polyvinylchloride) may show some chlorine evaporation. If temperature variation is high and continuous, at the end the plastic profile may give rise to evaporation. The vapour condenses on the surface of glass to give gray-brown colouration.

### 7.9.3 Quality Problems in Coated Glasses

Float glass production can be considered to have reached maturity. Every producer can more or less make all types with the exception of very thin and thick glasses; coloured glasses also remain challenging for some.

In the last few decades, research on coating glass has reached its peak. Their production has been successful and has added a multiple value to glass that has not seen before. For glasses to be coated more stringent quality standards have to be set and followed [87].

Coated glasses prevent heat losses from buildings while at the same time allow light and thermal radiation to enter. For this reason they save energy and are environmentally friendly. Pyrolitic coating was the first method used; the low-e glasses produced by this method were rather strong and widely used in architecture and automotive industry.

Glass coatings produced by vacuum sputtering (TCO, ITO) are used in solar modules, optical films and imaging elements. Their outstanding properties are that they are able to transmit light and are electrically conductive. Small scratches and defects which are like seeds do not negatively affect their function. But if the size of these defects becomes large enough then they cause problems. These can easily be examined under a microscope. If there is a thickness variation in the coating then some colouration may occur. It is possible to locate pinholes, scratches and colour variations. This will ensure that products sent to the customer are unquestionable. The information accumulated from the defects is used for necessary improvements in the process.

Glasses to be coated must be cleaned extremely well and controlled with stringent methods complemented with cameras and detectors.

## 7.10 Defects Encountered in Automotive Glasses

Automotive glasses are of the highest quality, where the glass producers do not feel at ease. It is a challenging area of float production, only some producers feel competent enough to produce automotive glass. There are two important points; one is to be able to produce good quality thin glass and the other is ideally those glasses produced for the front windscreen should be free of any distortions. Windscreen glasses are produced by lamination, again, these should not have any distortions [88,89].

Rear window glasses can if preferred be laminated or tempered. The rear and side glasses can be lower in quality compared to front screen. Heat treated glasses must also be of high quality, nevertheless the heating process itself can impart some distortions. In general, the glass is heated to a temperature of 650°C and is either moulded or allowed to form by gravity.

In this case the primary properties of float glass see noticeable deterioration. Distortions seen in the glasses are either due to stretching or thickness variations. During heating of the glass stretching occurs on the concave side, while on the convex side wrinkling may happen. The thickness of the glass must not change at the end of these operations.

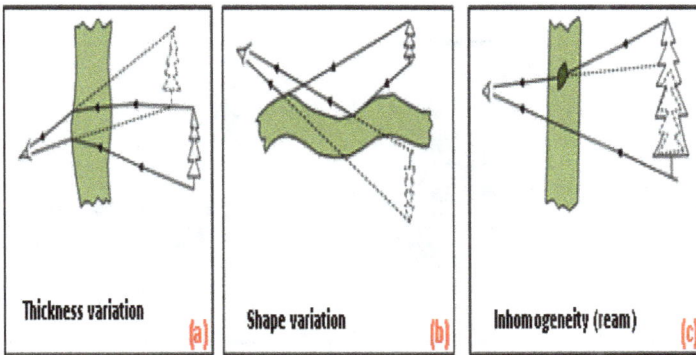

Thickness variation (a)    Shape variation (b)    Inhomogeneity (ream) (c)

Figure 83: Thickness and shape variations cause vision problems for the driver. Probably the most frequent and most severe is the effect of ream. Ream causes the same effect as thickness variation.

Especially in windscreen glasses **ream values** should not exceed 3–4. Otherwise repeated or double images may result. The driver can be disturbed by a **double image** and the distortion it causes, Figure 83, 84. The driver does not look at the glass, but looks through the glass.

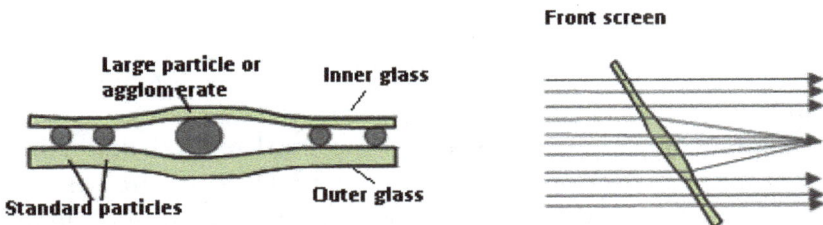

Front screen

Large particle or agglomerate    Inner glass

Standard particles    Outer glass

Figure 84: **Optical distortion effect** frequently seen in heat treated laminated glass.

**Zebra angles** of glasses must also be evaluated. The Zebra angles of thin glasses are generally lower at the sides, therefore glasses should be paired taking into account this point also, Figure 85. This variation in the angles is due to the cooling effect at the sides of the ribbon in the tin bath. The edges may be subjected to an edge lift effect (upward) at micron levels, Figure 67. It may be advisable therefore not to use the sides of the ribbon (right and left) in the windscreen, which will satisfy high value customers. There will be no problem using these glasses in the side windows.

Zebra angle

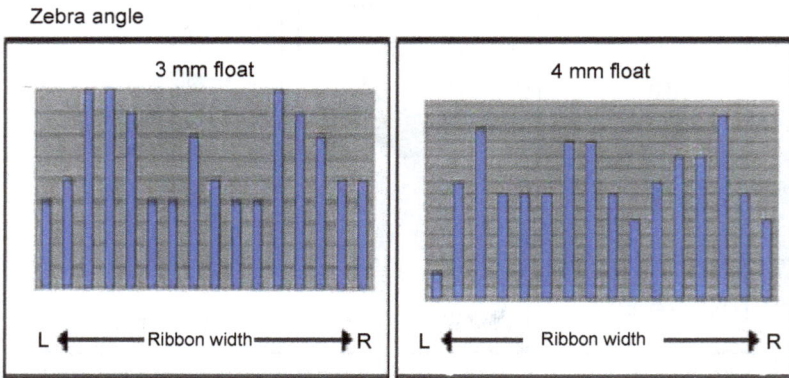

Figure 85: **Zebra angles** show more variations in thin glass than thicker glass. This property must be taken into account during processing.

Green glass to be used in automotive applications should not wait in stock for very long otherwise **scum formation** ($CaCO_2$) may happen on the bottom surface. This development is seen only in this type of glass. If there is a problem like this then it may be useful to wash the glass with acetic acid solution of pH=3.

If glass is to be heat treated it must be made absolutely clear that the tin side is placed on the outside/concave surface, otherwise **bloom** is likely to form. Detailed information about bloom was given before, Figure 46.

Laminated glasses may show some **local optical distortions**, which may be a serious quality problem [90,91], Figure 85. These local optical defects can only be found under special illuminating conditions. After marking the defects then the glass is cut to suitable size (be careful the glass may be heat treated). The PVB layer is separated at 240°C, and necessary investigations are made by using micro analytical techniques. The size of these particles may be 5–10 µm, enough to cause a disturbance. Electron microprobe analyses show that these particles are of the glass composition. Therefore, all parts of the plant may be a potential source, general cleanliness is important. Of special importance is that there should not be any glass dust in places where the glass is heat treated or laminated, even more effective these places may be isolated from the rest of the plant.

In the sides or near the edges of automotive glasses radio antenna and various sensors are imprinted by ceramic paints. The paints used for heating elements may contain silver. These paints are applied by serigraphy then heat treated. There should not be lead in the paints and they must be compatible with clips to be used. Organic paints are not resistant therefore they are not used in automotive glasses. The firing temperatures of organic paints is between 120–180°C, while that of ceramic paints is 800°C. This is why the chemical and mechanical resistance of organic paints are low.

# REFERENCES

1.  Smrcek, A. Composition of Industrial Glass. In *Fiberglass and Glass Technology*, Eds Frederick T. Wallenberger & Paul A. Bingham, Springer, 2010.
2.  Daneo, A.G. Falcone, R. & Hreglich, S. Effect of the Redox State on Container Glass Colour Stability. *Glass Technol.: Eur. J. Glass Sci. Technol. A*, 2009, **50** (3), 147–150.
3.  Aydin, E. An Approach to Correlate the $Al_2O_3$ Content with the Alkali Feldspar Used in Float Glass, *Eighth ESG Conf.*, Sunderland, England, 2006.
4.  Buscella, A. Glass Defects Caused by Raw Material Contamination. *Glass Mach. Plants Access.*, 2008, (1), 72–76.
5.  Aydin, E. & Arman, M. B. Catscratch Defects in Glass Containers and Tableware: A New Approach in Evaluation. *Int. Glass J.*, 2000, 106.
6.  Aydin, E. & Oran, M. The Behaviour of Cord Making Material in F/H Channels as Depicted by Physical and Mathematical Modelling, *Tenth ESG Conf.*, Magdeburg, Germany, 2010.
7.  Aydin, E. *Non-Classical Views about Traditional Glass Melting*, Brig, Switzerland, 2008.
8.  Aydin, E. Catscratches in Glass Products, Furnace Solutions 6, Stoke-On-Trent, England, 2011.
9.  Sesha Prakash, N. Inhomogeneities in Glass Cords and Striae. *Glass Mach. Plants Access.*, 2009, (20), 56–60.
10. Sesha Prakash, N. Control in Container Glass Making: A Critical Element. *Glass Int.*, Sept. 2009, 44–47.
11. Atkarskay, A. R. Effect of the Oxidation – Reduction Potential on the proneness of Glass to Form Bubbles. *Glass Ceram.*, 2010, **67** (3–4), 99–104.
12. Oran, M. İleri Cam Teknolojisi Notları, *Araştırma ve Teknolojik Geliştirme Başkanlığı*, Sisecam, 2004, p. 205–247

13. Aydin, E. Glass Raw Materials: Cost and Quality Aspects, *Proc. First Balkan Conference on Science and Technology*, Volos, Greece, 2000.

14. Johansen, J. R. Predicting Segregation of Bimodal Particle Mixtures using the Flow Properties of Bulk Solids. *Pharma. Technol. Eur.*, Publ. No. 0124, 1996.

15. Barnum, R. & Clement, S. Segregation During Batching. *Glass*, May 2005, 128.

16. Tang, P. & Puri, V. M. Methods of Minimising Segregation (A Review). *Partic. Sci. Technol.*, 2004, **22**, 321–337.

17. Carson, J. W., Royal, T. A. & Goodwill, D. J. 1986, Understanding and Eliminating Particle Segregation Problems. *Bulk Solids Handl.*, Feb. 1986, **6**, 139–144.

18. Martens, R., van Dick, F., Beerkens, R., van Kersbergen, M. & van Limpt, H. Carry-Over in Glass Furnaces, DGG Annual Meeting, DGG, Nürnberg, 2004.

19. Chopinet, M. H., Gouillart, E., Papin, S. & Toplis, M. J. Influence of Limestone Grain Size on Glass Homogeneity. *Glass Technol.: Eur. J. Glass Sci. Technol. A*, June 2010, **51** (3), 116–122.

20. Aydin, E. & Çebi, A. Decrepitation May Dictate your Choice of Limestone and Dolomite, *Int. Congr. Glass*, Kyoto, Japan, 2004.

21. Aydin, E., Oran, M., Gönül, T. & Ungan, A. Experimental and Numerical Investigation of Chromite Dissolution in Soda-Lime-Silica Melt, 18th Int. Congr. on Glass, San Francisco, USA, 1998.

22. Heymann, C. & Alt, K. Financial Aspects of Furnace Optimisation. *Glass Worldwide*, 2011, (35), 66–69.

23. Heymann, C. Financial Optimisation of Furnaces by Balancing Calculations. *Glass Int.*, March 2011, 15.

24. Houdaer, J. P. Les Principales Matieres Premieres Verrieres. *Verre*, Nov.–Dec. 1998, **4**, 17–24.

25. Simmingsköld, B. *Raw Materials for Glass Making*, Society of Glass Technology, 1997.

26. Hoyle, C. J. & Davis, D. H. Batch Charging: A key to Quality and Efficiency. *Glass Int.*, May 2010, 49–50.

27. West, M. Practical Guide for the Assessment of Glass Making Sands: Part 1 – Procedures of Sampling and Physical Testing. *Glass Technol.: Eur. J. Glass Sci. Technol. A*, 2010, **51** (2), 81–85.

28. Henderson, J. Defects from the Melting Process. *Int. Glass J.*, 2000, (106), 20–23.
29. Hoyle, C. J. & Davis, D. H. Benefit from Viewing the Dance of the Batch. *Glass Worldwide*, 2010, (30), 16–17.
30. Lehman, R. L. & Manring, W. H. Glass Batch Wetting with Water, *Batching in the Glass Industry*, 1981, p. 76–80.
31. FMC Soda Ash, 2000, Storage Operations and Technical Data http://www.fmcchemicals.com/division_alkalichemicals.asp
32. Litvin, V. I., Tokarev, V. D. & Yachevskii, A. V. Optimization of the Physico-Chemical Processes Occuring During the Preparation of Glass Mix and Evaluation of the Effect of the Mix Moisture Content on the Efficiency of the Glass-Making Process. *Glass Ceram.*, 2010, **67** (7–8), 244–248.
33. Worrell, E., Galitsky, C., Masanet, E. & Graus, W. *Energy Efficiency Improvement and Cost Saving Opportunities for the Glass Industry*, US Environmental Protection Agency, 2007.
34. Clark-Monks, C. & Parker, J. M. *Stones and Cords in Glass*, Society of Glass Technology 1980.
35. Bange, K., Müller, H. & Strubel, C. Characterisation of Defects in Glasses and Coatings on Glasses by Microanalytical Techniques. *Microchim. Acta*, 2000, **132**, 493–503.
36. Müller, H., Strubel, C. & Bange, K. Characterization and Identification of Local Defects in Glass. *Scannig*, 2000, **23**, 14–23.
37. Heitz, J.-L. Analyse et Controle Due Verre an Laboratoire. *Verre*, 2005, **11** (2), 8–14.
38. İzmirlioğlu, E. B., Yılmaz, Ş. & Aydın, E. Flat Cam Üretiminde Silika ve Banyo Hataları, Analitik Destek Hizmetleri Müdürlüğü Araştırma ve Teknoloji Genel Md. Yrd., Şişecam, 2009.
39. Özcan, A. Arman, M. B. Aydin, E. Siliceous Defects in Float Glass Production, Int. Symposium on Glass Problems, İstanbul, Turkey, 1996.
40. Beerkens, R. New Developments in Melting Technology and Energy Efficiency for Glass Production. *Int. Glass J.*, 2000, (106), 13–19.
41. Muschick, W. A Dynamic Simulation of Foam in Numerical Glass Tank Models, Schott Glass Central R&D, Mainz, 2001.
42. Aydin, E. Refractories and Priorities, Furnace Solutions 6,

Stoke-on-Trent, England, 2011.

43. Arman, M. B. & Aydin, E. The Evolution of Knots and Cords in Glass Products, 14th Int. Congress on Glass, New Delhi, India, 1986.

44. Manfredo, L. J. & McNally, R. N. Solubility of Refractory Oxides in Soda-Line Glass. *Commun. Am. Ceram. Soc.*, p. C155–C158; *J. Mater. Sci.*, 1984, **19**, 1272–1276.

45. Ross, P. C. Exudation of Fused Cast Refractories. *Glass Int.*, October 2007, 6.

46. Gaskell, L. Cord Dispersal System Reduces Catscratches on Glass. *Glass Int.*, Nov. 2011, 17–19.

47. Sesha Prakash, N. Causing a Stir in Glass Making. *Glass Int.*, Nov. 2011, 21–26.

48. Sesha Prakash (Personal Communication).

49. Feng, Z., Li, D., Qin, G. & Liu, S. Study of the Float Glass Melting Process: Combining Fluid Dynamics Simulation and Glass Homogeneity Inspection. *J. Am. Ceram. Soc.*, 2008, **91** (10), 3229–3234.

50. Feng, Z., Li, D., Qin, G. & Liu, S. Effect of the Flow Pattern in a Float Glass Furnace on Glass Quality: Calculations and Experimental Evaluation of On-Site Samples. *J. Am. Ceram. Soc.*, 2009, **92** (12), 3098–3100.

51. Guloyan, Yu. A. Certain Particularities of Transport Phenomena in Glass Making Furnaces. *Glass Ceram.*, 2009, **66** (5–6), 197–201.

52. Dzyuzer, V. Ya. Perfecting The Technical Characteristics of Glass-Making Furnaces (Review). *Glass Ceram.*, 2008, **66** (7–8), 217–226.

53. Kucera, J., Krobot, T. & Skrivan, M. New Physical Modelling İnfluences Glass Homogeneity. *Glass Int.*, April 2010, 31–34.

54. Falleroni, C. A. & Edge, C. K. Float Glass Technology: The Bath Atmosphere System, PPG Industries Inc, 1996, p. 61–70.

55. Edge, C. K. Float Glass Manufacture: The Control of Tin Flows, PPG Industries Inc, 1995, p. 31–42.

56. Frischat, G. H. Tin Ions in Float Glass Causes Anomalies. *C.R. Chimie*, 2002, **5**, 759–763.

57. Frantz, H. Ion Exchange and Redox Reactions in the Float Bath. *Glastech. Ber. Glass Sci. Technol.*, 1995, **68**, C1, 15–19.

58. Verita, M., Geotti-Biancini, F., Guadagnino, E., Stella, A., Pantano, C. G. & Paulson, T. Chemical Characterisation of the Bottom Side of Green Float Glasses. *Glasstech. Ber. Glass Sci. Technol.*, 1995, **68** (C1), 251–258.

59. Williams, K. F. E., Thomas, M. F., Greengrass, J. & Bradshaw J. M. The Effect on Tin on Some Physical Properties of the Bottom Surface of Float Glass and the Origin of Bloom. *Glass Technol.*, 1999, **40** (4), 103–107.

60. Laimbock, P. R. & Beerkens, R. G. C. Oxygen Sensor for Float Production Lines. *Am. Ceram. Soc. Bull.*, May 2006, **85**, 33–36.

61. Balbo, M. Process and Product Quality in the Production of Float Glass. *Int. Glass J.*, 2003, (128), 26.

62. Liu, D. R., Park, J. S., Benoit, R. E. & Jackson, F. D. Chemical Analysis of Ream Defect in Float Glass. *X-Ray Spectr..*, 1992, **21**, 293–298.

63. Sesha Prakash, N. Rollers Improve Flat Glass Annealing. *Glass Int.*, February 2011, 29–33.

64. Douglas, R. W. & Isard, J. O. Action of Water and of Sulfur Dioxide on Glass Surfaces. *J. Soc. Glass Technol.*, 1949, **33**, 289–335.

65. Arman, M. B., Özcan, A., Demirli, Ş. & Orhon, M. Carbonaceous Minerals as Corrosion Products on the Surfaces of Bottles, Int. Congress on Glass, CD-ROM, Edinburgh, Scotland, 2001.

66. Aydin, E. Perlite Message in a Bottle. *Ind. Miner.*, April 2006, 73–77.

67. Ross, C. P. Amber Glass – 40 years of Lessons Learned, 66th Conference on Glass Problems, 2006, p. 129–136.

68. Jain, H. & Sharma, A. Surfaces of Common Commercial Glass Products: Inconspicuous Defects. *Am. Ceram. Soc. Bull.*, 2005, **84** (7), 33–37.

69. Sökmen, İ. Bulaşık Makinesinde Yıkama ile Cam Yüzeyinde Ortaya Çıkan Damar Hatası, *Teknik Bülten*, 2007, **C. 36** (4), 41–47. Araştırma ve Teknoloji Genel Md. Yrd., Şişecam.

70. İzmirlioğlu, E. B. & Sökmen, İ. B. Yıkama Süngerlerinin Cam Ev Eşyalarına Etkisi. *Teknik Bülten*, 2007, **C. 36** (2). Araştırma ve Teknoloji Genel Md. Yrd. Şişecam.

71. Glaser, H. J. *Large Area Glass Coating*, Ed. Von Ardenne Anlagentechnik GmbH, Dresden, Germany, 2005.

72. Falcone, R., Licenziati, F., Orsega, E. F. & Verita, M. The Dependence of the Weathering of Soda-Lime-Silica Glass on Environmental Parameters: A Preliminary Investigation. *Glass Technol.: Eur. J. Glass Sci. Technol. A*, 2011, **52** (1), 23–39.

73. Adams, P. B. Glass Corrosion. *J. Non-Cryst. Solids*, 1984, **67**, 193–205.

74. Sesha Prakash, N. Handling, Storage and Transport of Flat Glass. *Glass Int.*, Oct 2011, 19–22.

75. Guloyan, Yu. A. & Shelomentseva, V. F. A Study of Physicochemical Phenomena in Evaluation of Chemical Corrosion in Glass. *Glass Ceram.*, 2000, **57** (7–8), 267–271.

76. Guloyan, Yu. A. Serviciability of Glass and Glassware (Review). *Glass Ceram.*, 2008, **65** (5–6), 177–186.

77. Walten H. V. and Adams, P. B. Effects of Humidity on the Weathering of Glass. *J. Non-Cryst. Solids*, 1975, **19**, 183–199.

78. Radlein, F. Brokmann, U. Hesse, A. Common Features of Glass Surface Degradation, Glass Performance Days, 2009, p.202–204.

79. Schaut, R. A. & Pantano C. G. Acid Interleave Coatings Inhibit Float-Glass Weathering, Corrosion. *Am. Ceram Soc. Bull.*, 2005, **84**, (10), 44–45.

80. Xu Chao, Cheng Jinshu, Guo Liping, The Mixed Alkali Effect and The Resistance of Float Glasses to Water, Proc. XVII Int. Congress on Glass, 1995, Vol. 3, p. 114–120.

81. Shen Zuwu, Liu Qiming, Meng Li & Wang Hongcheng. Effect of Additional Applied Thermal/Electric Field on the Float-glass Surfaces. *J. Wuhan Univ. Tech.-Mater. Sci. Ed.*, 2009, **24**, (2), 308–311.

82. Sharagov, V. Dealkalisation of Glass Surfaces by Acid Gasses in the Electric Field. *Nonconventional Technol. Rev.*, 2007, (3), 97–100.

83. Falcone, R., Licenziati, F., Orsega, E. F. & Verità, M. The Dependence of Weathering of Soda–Lime–Silica Glass on Environmental Parameters. *Glass Technol.: Eur. J. Glass Sci. Technol. A*, 2011, **52** (1), 23–29.

84. Josey, B. Glass for Buildings – is it Crystal Clear? *Struct. Survey*, 1997, **15** (1), 15–20.

85. Specifiers Guide to Architectural Glass, 2005 Edition,

Glass Association of North America (GANA), Flat Glass Manufacturing Division, p. 1–17.

86. Kasper, A. Nickelsulfid im Vorgespannten Glass. *Glaswelt*, 2000, **3**, 67–71.

87. Oerley, H. & Bauereiss, U. Optical Quality Assurance and Process Control of Coated Glass. *Glass Worldwide*, 2011, (33), 38–40.

88. Woodward, A. Looking into Automative Glass Technology. *Glass Ind. Dev. Int.*, 2002, 77–79.

89. Nagel, J. & Pingel, U. Measurement and Inspection on Float Glass and Automotive Glazings. *Verre*, 2005, **11** (2), 24–27.

90. Pişirici, Ö., Suner, Ç. & Akçakaya, R. Derin Bombeli Ön Camlarda Oluşan Optik Problemi ve Çözümü, 21. Cam Sempozyumu, 2006, s. 90–95.

91. Yuneva E. V., Gorokhovskii, A. V. & Kaplina, T. V. Improving the Optical Properties of Float Glass. *Glass Ceram.*, 2006, **63** (11–12), 359–361.

# APPENDIX 1 Minerals Found in Glass Raw Materials

| Minerals | Mineral formula | Key components | Chemical composition (wt%) | Melting or Decomposition (°C) | Density |
|---|---|---|---|---|---|
| Albite | $Na_2O.3Al_2O_3.6SiO_2$ | $Na_2O$, $Al_2O_3$ | $Na_2O$ 11.8; $Al_2O_3$ 19.4; $SiO_2$ 68.8 | M 110 | 2.61–2.64 |
| Alunite | $K_2O.Al_2O_3.4SiO_2.6H_2O$ | | $K_2O$ 11.4; $Al_2O_3$ 37.0; $SiO_2$ 38.6; $H_2O$ 13.0 | D 800 | 2.75 |
| Anorthite | $CaO.Al_2O_3.2SiO_2$ | $CaO$, $Al_2O_3$ | $CaO$ 20.2; $Al_2O_3$ 36.6; $SiO_2$ 43.2 | M 1550 | 2.77 |
| Brucite | $Mg(OH)_2$ | | $Mg$ 69.1; $H_2O$ 30.9 | D 400 M2830 | 2.4 |
| Calcite | $CaCO_3$ | $CaO$ | $Ca$ 56.0; $CO_2$ 44.0 | D 900 M2570 | 2.71 |
| Chromite | $FeO.Cr_2O_3$ | $Cr_2O_3$ | $FeO$ 32.1; $Cr_2O_3$ 67.9 | M 2180 | 4.3–4.57 |
| Cordierite | $2MgO.2Al_2O_3.5SiO_2$ | | $MgO$ 13.8; $Al_2O_3$ 34.9; $SiO_2$ 51.3 | M 1460 | 2.57–2.66 |
| Diaspore | $Al_2O_3.H_2O$ | | $Al_2O_3$ 85.5; $H_2O$ 15.0 | M 2050 | 3.3–3.5 |
| Diopside | $CaO.MgO.2SiO_2$ | | $CaO$ 25.9; $MgO$ 18.6; $SiO_2$ 55.5 | M 1391 | 3.20–3.38 |
| Dolomite | $MgCO_3.CaCO_3$ | $MgO$, $CaO$ | $MgO$ 21.9; $CaO$ 30.4; $CO_2$ 47.7 | D 800 (in air) | – |

| Minerals | Mineral formula | Chemical composition (wt%) | | | Melting or Decomposition (°C) | Density |
|---|---|---|---|---|---|---|
| Fayalite | $2FeO.SiO_2$ | FeO 70.5 | SiO₂ 29.5 | | M 1205 | 3.91–4.54 |
| Forsterite | $2MgO.SiO_2$ | MgO 57.2 | SiO₂ 42.8 | | M 1896 | 3.19–3.33 |
| Kaliophite | $K_2O.Al_2O_3.2SiO_2$ | K₂O 29.8 | Al₂O₃ 32.2 | SiO₂ 38.0 | – | 2.61 |
| Kyanite | $Al_2O_3.2SiO_2.2H_2O$ | Al₂O₃ 39.5 | SiO₂ 46.5 | H₂O 14.0 | – | 2.60–2.63 |
| Leucite | $K_2O.Al_2O_3.4SiO_2$ | K₂O 21.6 | Al₂O₃ 23.3 | SiO₂ 55.1 | M 1750 | 4.6 |
| Magnesite | $MgCO_3$ | MgO 47.8 | CO₂ 52.2 | | – | 2.95–3.2 |
| Magnetite | $FeO.Fe_2O_3$ | FeO 31.1 | Fe₂O₃ 68.9 | | M 1591 | 4.96–5.20 |
| Monticellite | $CaO.MgO.SiO_2$ | CaO 35.9 | MgO 25.7 | SiO₂ 38.4 | M 1610 | 3.2 |
| Mullite | $Al_2O_3.SiO_2$ | Al₂O₃ 71.8 | SiO₂ 28.2 | | M 1810 | 3.03 |

| Minerals | Mineral formula | Chemical composition (wt%) | | | | | | | | Melting or Decomposition (°C) | Density |
|---|---|---|---|---|---|---|---|---|---|---|---|
| Nephelite | $Na_2O.Al_2O_3.SiO_2$ ($Na_2O$ $Al_2O_3$) | $Na_2O$ | 21·8 | $Al_2O_3$ | 35·9 | $SiO_2$ | 42·3 | | | M 1526 | 2·55–2·65 |
| Orthoclase | $K_2O.Al_2O_3.6SiO_2$ ($K_2O$ $Al_2O_3$) | $K_2O$ | 16·9 | $Al_2O_3$ | 18·3 | $SiO_2$ | 64·8 | | | M 1150 | 2·6 |
| Sericite | $K_2O.3Al_2O_3.6SiO_2.2H_2O$ | $K_2O$ | 11·8 | $Al_2O_3$ | 38·5 | $SiO_2$ | 45·2 | $H_2O$ | 4·5 | D 950 | – |
| Serpentine | $3MgO.2SiO_2.2H_2O$ | $MgO$ | 43·6 | $SiO_2$ | 43·4 | $H_2O$ | 13·0 | | | D 500–700 | 2·6–3·0 |
| Silimanite | $Al_2O_3.SiO_2$ | $Al_2O_3$ | 62·9 | $SiO_2$ | 37·1 | | | | | M1810 | 3·23–3·25 |
| Spinel | $MgO.Al_2O_3$ | $MgO$ | 28·3 | $Al_2O_3$ | 71·7 | | | | | M 2135 | 3·5–4·1 |
| Wollastonite | $CaO.SiO_2$ ($CaO$) | $CaO$ | 48·3 | $SiO_2$ | 51·7 | | | | | M 1540 | 2·8–2·9 |
| Zircon | $ZrO_2.SiO_2$ ($ZrO_2$) | $ZrO_2$ | 67·2 | $SiO_2$ | 32·8 | | | | | M 1675 | 4·7 |

Notes: The minerals from which oxides of glass making are extracted are shown with different colours eg, blue colour indicates the minerals from which alkali oxides ($R_2O$) and $Al_2O_3$ are derived.

# APPENDIX 2 Accidental Mixing of Glass Raw Materials

| MIXED RAW MATERIAL \ XRF ANALYSES OF INDICATIVE OXIDES | Sand | Limestone | Dolomite | Feldspar | Soda | Sulphate |
|---|---|---|---|---|---|---|
| **Sand** |  | $SiO_2$ / $CaO$ | $SiO_2$ / $CaO, MgO$ | $SiO_2$ / $Na_2O, Al_2O_3$ | $SiO_2$ / $Na_2O$ | $SiO_2$ / $SO_3, Na_2O$ |
| **Limestone** | $CaO$ / $SiO_2$ |  | $CaO$ / $CaO, MgO$ | $CaO$ / $Na_2O, Al_2O_3$ | $CaO$ / $Na_2O$ | $CaO$ / $SO_3, Na_2O$ |
| **Dolomite** | $CaO, MgO$ / $SiO_2$ | $CaO, MgO$ / $CaO$ |  | $CaO, MgO$ / $Na_2O, Al_2O_3$ | $CaO, MgO$ / $Na_2O$ | $CaO, MgO$ / $SO_3, Na_2O$ |
| **Feldspar** | $Na_2O, Al_2O_3$ / $SiO_2$ | $Na_2O, Al_2O_3$ / $CaO$ | $Na_2O, Al_2O_3$ / $CaO, MgO$ |  | $Na_2O, Al_2O_3$ / $Na_2O$ | $Na_2O, Al_2O_3$ / $SO_3, Na_2O$ |
| **Soda** | $Na_2O$ / $SiO_2$ | $Na_2O$ / $CaO$ | $Na_2O$ / $CaO, MgO$ | $Na_2O$ / $Na_2O, Al_2O_3$ |  | $Na_2O$ / $SO_3, Na_2O$ |
| **Sulphate** | $Na_2O, SO_3$ / $SiO_2$ | $Na_2O, SO_3$ / $CaO$ | $Na_2O, SO_3$ / $CaO, MgO$ | $Na_2O, SO_3$ / $Na_2O, Al_2O_3$ | $Na_2O, SO_3$ / $Na_2O$ |  |

**Special Note:**
In the glass industry it is quite likely that some raw materials may accidentally mix with one another. The volume may be high or low.

When there is a deviation from chemical composition immediate action must be taken. If an XRF instrument is available or easy to access then it is relatively easy to find out which raw material is mixed. The following points are critical:

- Decide which grain size range to use, this will save time,
- Prepare a press sample for XRF, then count and note the count levels of elements, note the table above. A raw material on the left column is mixed with any of the materials on the right. For example if feldspar is mixed, then we must look for the presence of ($Na_2O, Al_2O_3$), If dolomite is the one that is mixed then search for ($CaO, MgO$); if sulphate is mixed then check ($Na_2O$, $SO_3$). The oxides coloured in blue are the indicator oxides of the material mixed.
- Using classical methods of wet chemistry is slow and time consuming. However, using a suitable acid (HCl) to dissolve the soluble components (carbonates, sulphates) is good way of reducing the size of sample to be investigated.

# Biography

**Dr Eşref Aydin** (71) was born in Turkey. He completed his higher education (BSc and MPhil) at University College London on a grant given by the Geological Survey (MTA) of Turkey. He accomplished his PhD externally at the Geology Department of University of Istanbul. After working at MTA for five years, he joined R&D of Şişecam, the world renowned glass producing company.

He specialised in refractories, glass raw materials, glass defects and glass quality at large, having been involved in problems of several glass types. He acted as the manager of Analytical Services for 20 years. During this time he has taken an active role in the activities of International Commission of Glass (ICG). He was the chairman of Refractories Committee (TC 11) for seven years. During this time he has contributed to many conferences as a speaker or author.

He was nade a Fellow of the Society of Glass Technology in 2010. He wrote a book (in Turkish) titled "Problematic Areas in Glass Production" in 2012. After 32 years working in Şişecam he was retired in 2012. He lectured on part time basis at the Technical University of Istanbul for four years. He is still acting as a consultant.

# Subject Index

www.ingramcontent.com/pod-product-compliance
Lightning Source LLC
Chambersburg PA
CBHW070719220326
41598CB00024BA/3232